Leckie × Leckie
Scotland's leading educational publishers

Practice Papers for SQA Exams

National 5

Biology

001/29012018

10 9 8 7 6 5 4 3 2 1

ISBN 9780008281670

Published by
Leckie & Leckie Ltd
An imprint of HarperCollins*Publishers*
Westerhill Road, Bishopbriggs, Glasgow, G64 2QT
T: 0844 576 8126   F: 0844 576 8131
leckieandleckie@harpercollins.co.uk

www.leckieandleckie.co.uk

Special thanks to
QBS (layout and illustration); Ink Tank (cover design);
Jill Laidlaw (copy-edit); Paul Sensecall (proofreading);
Rona Gloag (proofreading); Dylan Hamilton (proofreading)

A CIP Catalogue record for this book is available from the British Library.

Acknowledgements
Whilst every effort has been made to trace the copyright holders, in cases where this has been unsuccessful, or if any have inadvertently been overlooked, the Publishers would gladly receive any information enabling them to rectify any error or omission at the first opportunity.

Printed in Italy by Grafica Veneta SpA

Introduction

The three papers included in this book are designed to provide practice in the National 5 Biology course assessment question paper (the examination), which is worth 80% of the final grade for this course.

Together, the three papers give overall and comprehensive coverage of the assessment of **knowledge and its application** as well as the **skills of scientific inquiry** needed to pass National 5 Biology. The **Key Area Index** grid on page 5 shows the pattern of coverage of the knowledge in the key areas and the skills across the three papers.

We recommend that candidates download a copy of the Course Specification from the SQA website at www.sqa.org.uk. Print pages 25–47, which summarise the knowledge and skills that will be tested.

Design of the papers

Each paper has been carefully assembled to be very similar to a typical National 5 question paper. Each paper has 100 marks and is divided into two sections.

- **Section 1** – objective test, which contains 25 multiple choice items worth 1 mark each, totalling 25 marks.
- **Section 2** – paper 2, which contains structured and extended response questions worth 1 to 4 marks each, totalling 75 marks.

In each paper, the marks are distributed evenly across all three component areas of the course, and the majority of the marks are for the demonstration and application of knowledge. The other marks are for the application of skills of scientific inquiry. We have included features of the national papers such as including a scientific literacy question, offering choice in some questions and building in opportunities for candidates to suggest adjustments to investigation and experimental designs.

Most questions in each paper are set at the standard of Grade C, but there are also more difficult questions set at the standard for Grade A. We have attempted to construct each paper to represent the typical range of demand in a National 5 Biology paper.

Using the papers

Each paper can be attempted as a whole, or groups of questions on a particular topic or skill area can be tackled – use the **Key Area Index** grid to find related groups of questions. In the grid, questions may appear twice if they cover more than one skill area. Use the 'Date completed' column to keep a record of your progress.

We recommend working between attempting the questions and studying their expected answers.

You will need a **pen**, a **sharp pencil**, **a clear plastic ruler** and a **calculator** for the best results. A couple of different **coloured highlighters** could also be handy.

Expected answers

The expected answers on pages 101–125 give national standard answers but, occasionally, there may be other acceptable answers. The answers have Top Tips provided alongside but don't feel you need to use them all!

The Top Tips include hints on the biology itself as well as some memory ideas, a focus on traditionally difficult areas, advice on the wording of answers and notes of commonly made errors.

Grading

The three papers are designed to be equally demanding and to reflect the national standard of a typical SQA paper. Each paper has 100 marks – if you score 50 marks, that's a C pass. You will need about 60 marks for a B pass and about 70 marks for an A. These figures are a rough guide only.

Timing

If you are attempting a full paper, limit yourself to **two hours and thirty minutes** to complete it. Get someone to time you! We recommend no more than 30 minutes for **Section 1** and the remainder of the time for **Section 2**.

If you are tackling blocks of questions, give yourself about a minute and a half per mark; for example, 10 marks of questions should take no longer than 15 minutes.

Good luck!

Topic index

Skill tested	Key area	Practice paper questions Section 1 – Objective test Section 2 – Paper 2						Date completed
		Paper A		*Paper B*		*Paper C*		
		Section 1	Section 2	Section 1	Section 2	Section 1	Section 2	
Area 1: Cell biology Demonstrating and applying knowledge	1. Cell Structure	1, 3		1			1	
	2. Transport across cell membranes	6	1	3, 4	1	1, 2	2a, bii	
	3. DNA and the production of proteins	8			3	4		
	4. Proteins		2	5, 6				
	5. Genetic engineering	9		7			3	
	6. Respiration		3b, 4	8		6, 7		
Area 2: Multicellular organisms Demonstrating and applying knowledge	1. Producing new cells	7, 9, 10, 11		10	2	8, 9	4d, 5	
	2. Control and communication	12, 13	5		7	10, 11	6	
	3. Reproduction	14		13	5	12		
	4. Variation and inheritance		6	11, 12	8c	13	7c, 8	
	5. Transport systems - plants	16	7	14, 15	10b, c, d	14, 15		
	6. Transport systems – animals		9		9, 16b, c		9a	
	7. Absorption of materials		8a	16		16		

Skill tested	Key area	Practice paper questions Section 1 – Objective test Section 2 – Paper 2						Date completed
		Paper A		*Paper B*		*Paper C*		
		Section 1	Section 2	Section 1	Section 2	Section 1	Section 2	
Area 3: Life on earth Demonstrating and applying knowledge	1. Ecosystems	18, 19		19		17, 20	11	
	2. Distribution of organisms	20	10a, d	17	11a	18	12c	
	3. Photosynthesis		13a	22		21	13ai, c	
	4. Energy in ecosystems	21		18		19	16	
	5. Food production	22	11		13c, d	24		
	6. Evolution of species	25	14	23	14		15	
Skills of scientific inquiry	1. Planning	24	4c, 13bii, iii	9, 21	4a, b, 6e		4c, 7a, 10, 13aii	
	2. Selecting	2	8bi, 10c, 12		4c, 8a, b, 10a, 11b, c, d, 12b, 13a, 15b, 16d	22, 23	7b, 13bi, 14a, b, d, e	
	3. Presenting		3a		6a, 12a		2bi	
	4. Processing	4, 15, 17, 23	8bii, 10b	2, 24	6b, 13b, 16a	25	4a, 9b, 13bii, 14c	
	5. Predicting		3d		15c		2biii	
	6. Concluding	5	13bi	20, 25	6c, 15a	3, 5	4b	
	7. Evaluating		3c		4d, 6d, 12b		12a, b	

Practice Paper A

SECTION 1 ANSWER GRID

Mark the correct answer as shown ✓

	A	B	C	D
1	○	○	○	○
2	○	○	○	○
3	○	○	○	○
4	○	○	○	○
5	○	○	○	○
6	○	○	○	○
7	○	○	○	○
8	○	○	○	○
9	○	○	○	○
10	○	○	○	○
11	○	○	○	○
12	○	○	○	○
13	○	○	○	○
14	○	○	○	○
15	○	○	○	○
16	○	○	○	○
17	○	○	○	○
18	○	○	○	○
19	○	○	○	○
20	○	○	○	○
21	○	○	○	○
22	○	○	○	○
23	○	○	○	○
24	○	○	○	○
25	○	○	○	○

N5

N5 Biology

Practice Papers for SQA Exams

Practice Paper A
Section 1

Fill in these boxes and read what is printed below.

Full name of centre

Town

Forename(s)

Surname

Try to answer ALL of the questions in the time allowed.

You have 2 hours and 30 minutes to complete this paper.

Write your answers in the spaces provided, including all of your working.

Leckie × Leckie
Scotland's leading educational publishers

SECTION 1 – 25 marks
Attempt ALL questions

1. The diagram below shows some structures present in a mesophyll cell from a green plant.

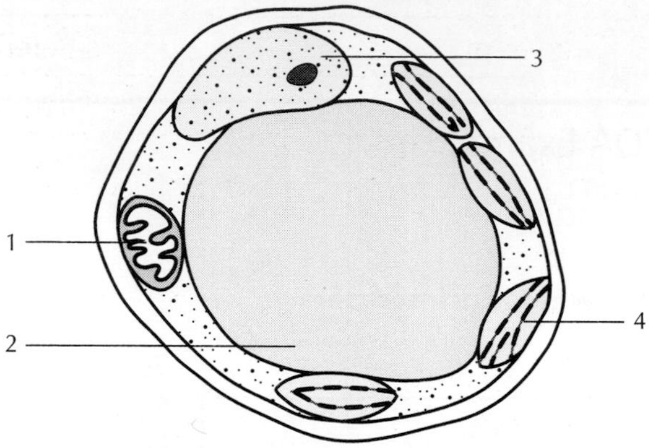

Which line in the table below identifies correctly the structures in the cell that carry out photosynthesis and contain genetic information?

	Carry out photosynthesis	*Contain genetic information*
A	1	2
B	4	3
C	1	3
D	4	2

2. The histogram below shows the number of cells of different lengths in a sample of onion epidermis.

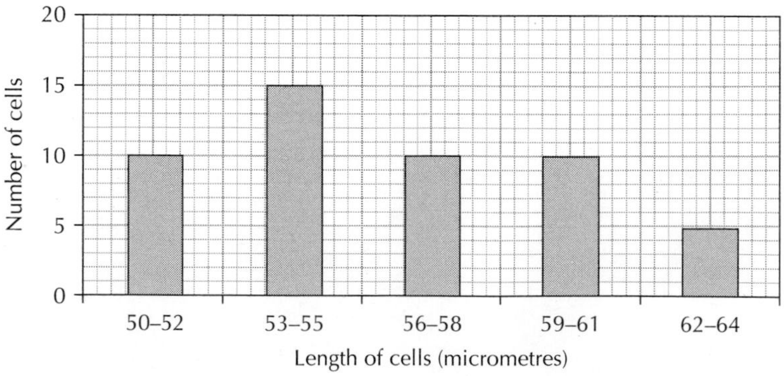

What percentage of the cells in the sample have a length greater than 58 micrometres?

A 15%

B 25%

C 30%

D 50%

3. Which line in the table below compares cell walls of plant and fungal cells?

	Chemical composition	***Structure***
A	same	same
B	different	different
C	same	different
D	different	same

4. The diagram below represents a bacterial cell as viewed under a microscope set to magnify 500 times.

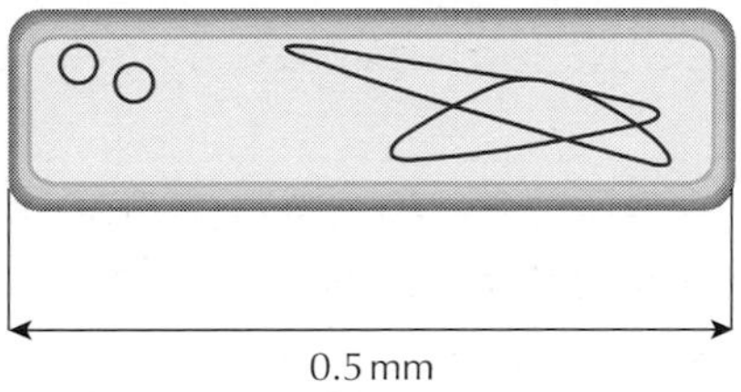

How many cells of this size would fit end to end into a millimetre?

A 10

B 100

C 500

D 1000

5. 50 mm strips of potato tissue were placed into each of three sucrose solutions P, Q and R of different concentrations and left at room temperature. After 1 hour the strips of tissue were re-measured and the results are shown in the table below.

Sucrose solution	***Length of potato tissue strip after 1 hour* (mm)**
P	50
Q	47
R	52

Which of the following conclusions based on these results is valid?

A Solution P had a lower concentration of sucrose than the potato cell sap

B Solution Q had a higher concentration of sucrose than the potato cell sap

C Solution R had a higher concentration of sucrose than the potato cell sap

D Solutions P, Q and R had the same concentration as the potato cell sap.

6. In active transport, molecules are moved by membrane

A proteins against the concentration gradient

B lipids down the concentration gradient

C lipids against the concentration gradient

D proteins down the concentration gradient.

7. The diagram below shows a stage in mitosis in a plant cell.

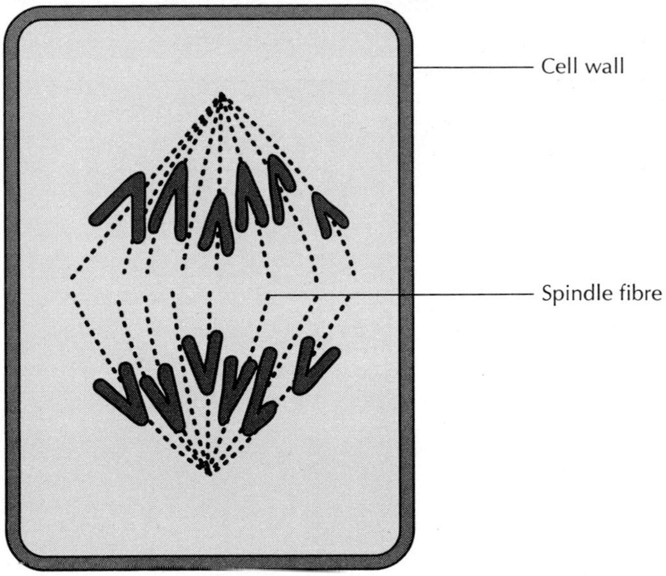

Which of the following best describes the chromosomes at the stage of mitosis shown?

The chromosomes have

A become visible as pairs of identical chromatids

B aligned at the equator of the spindle

C gathered at opposite poles of the spindle

D been pulled apart by spindle fibres.

8. The diagram below represents a short piece of a DNA molecule.

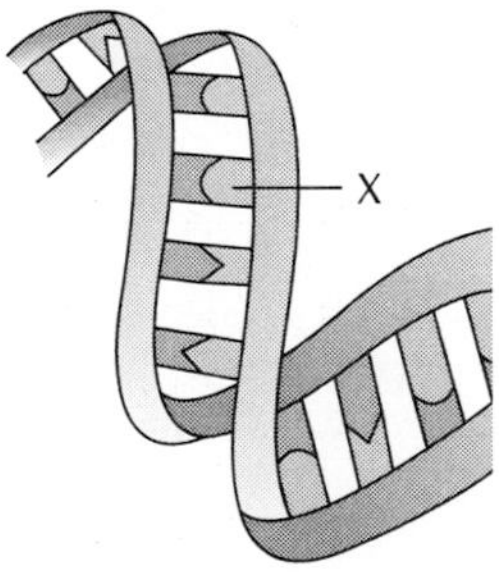

Which part of the DNA molecule is shown at X?

A Sugar

B Base

C Gene

D Amino acid.

9. The diagram below shows a genetically modified bacterial cell that contains a human gene.

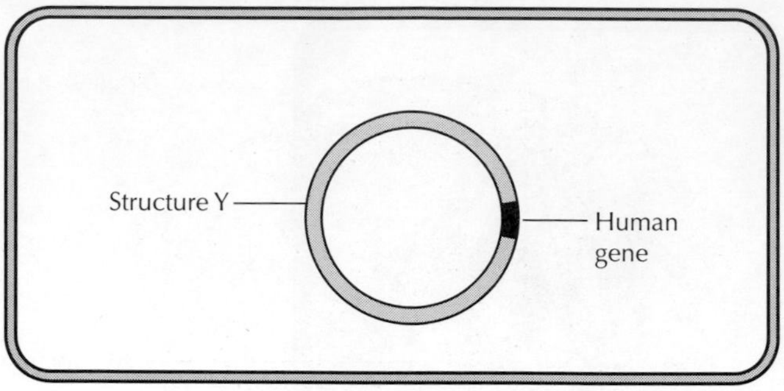

Structure Y, which contains the human gene, is

A the nucleus

B a chromosome

C a ribosome

D a plasmid.

10. A group of similar cells working together to perform the same function is called

A an organism

B a system

C an organ

D a tissue.

11. Which of the following statements is **false** in relation to stem cells?

Stem cells

A are found in animal embryos

B can undergo cell division

C develop into gametes

D can self-renew.

12. The diagram below shows a vertical section through the human brain.

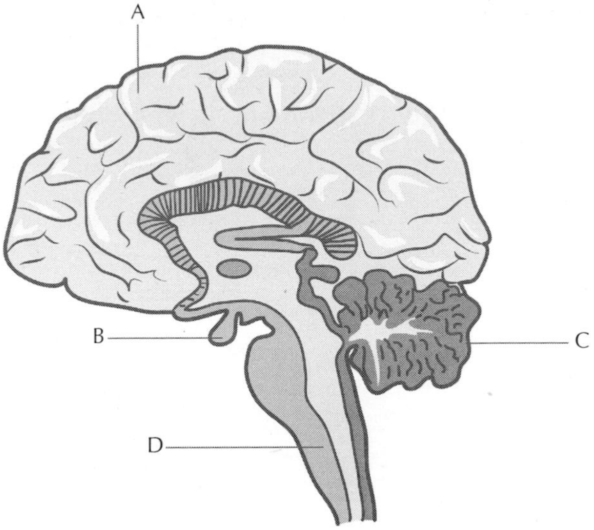

Which letter indicates the site of memory storage and reasoning?

13. Which organ contains target tissues that respond to insulin?

A Small intestine

B Pancreas

C Liver

D Brain.

14. Which line in the table below shows correctly the chromosome complements of the mammalian cells listed?

	Mammalian cell		
	muscle cell	***gamete***	***zygote***
A	diploid	haploid	haploid
B	diploid	haploid	diploid
C	haploid	diploid	diploid
D	haploid	diploid	haploid

15. The cardiac output from the heart is calculated using the equation shown below.

cardiac output (litres per min) = volume of blood pumped per beat (cm^3) × heart rate (beats per minute)

A hospital patient had a heart rate of 80 beats per minute and a cardiac output of 4 litres per minute.

What is the volume of blood pumped per beat?

A 5 cm^3

B 20 cm^3

C 50 cm^3

D 320 cm^3

Questions 16 and 17 refer to the following information.

A weight potometer was set up to compare the transpiration rates of a plant in different sets of environmental conditions.

The graph below shows the results of two experiments in which the environmental conditions were altered.

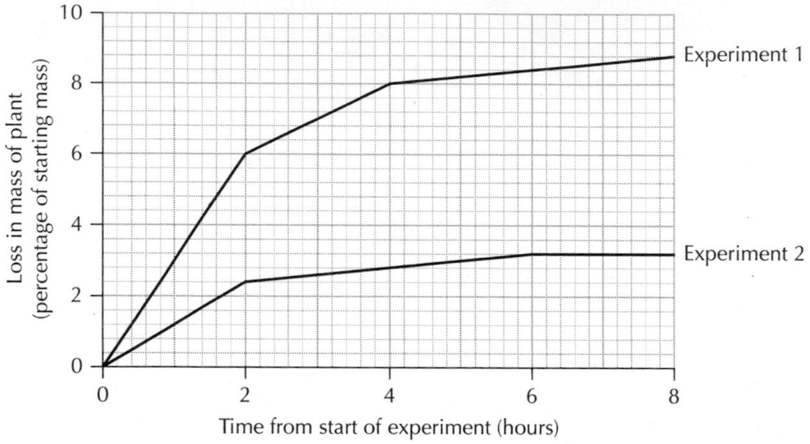

16. Which line in the table shows the possible conditions in Experiments 1 and 2 assuming all other conditions were kept constant?

	Experiment 1		***Experiment 2***	
	temperature	***humidity***	***temperature***	***humidity***
A	high	low	low	high
B	high	low	high	low
C	low	high	high	low
D	low	high	low	high

17. In Experiment 1, the plant had a mass of 600g at the start of the experiment.

What was its mass after 4 hours?

A 547.2g

B 552.0g

C 583.2g

D 592.0g

18. The total variety of all living organisms on Earth is described as its

A habitat

B biodiversity

C ecosystem

D population.

19. Which of the following statements is **true**?

The community of a Scottish moorland ecosystem consists of all the

A plant species present

B plant species present and the non-living environment

C plant and animal species present and the non-living environment

D plant and animal species present.

20. Which of the following factors are **both** biotic?

A Predation and temperature

B Temperature and pH

C pH and grazing

D Grazing and predation.

21. In the food chain below the plant plankton contains 100 000 units of energy from photosynthesis.

plant plankton → **animal plankton** → **small fish** → **predatory fish**

If 90% of the energy available at a food chain level is lost between levels, how many units of energy will be found in the predatory fish in the food chain above?

A 10 000

B 1000

C 100

D 10.

22. Which of these substances is absorbed from soil by plants and used in the synthesis of proteins?

A Amino acids

B Nitrates

C Sugars

D Water.

Questions 23 and 24 refer to the following information.

During a survey of the distribution of limpets on a rocky seashore, a number of quadrat samples were taken along a transect line between the high and low tide marks.

The numbers in the diagram below indicate the numbers of limpets in each quadrat.

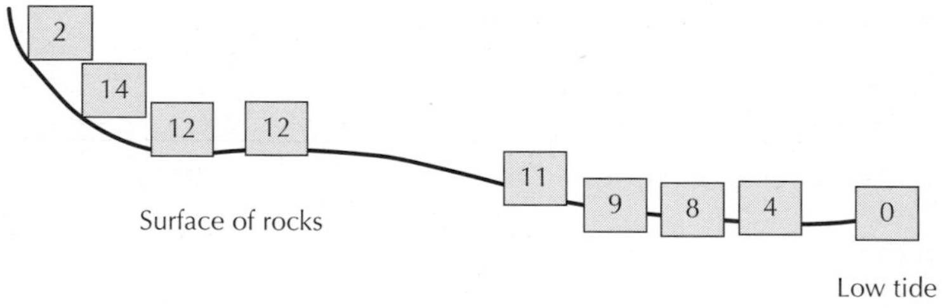

23. What is the average number of limpets per quadrat?

A 8

B 9

C 11

D 12

24. Which of the following is a precaution needed to make the results of the survey more valid?

A Place quadrats randomly

B Use exactly ten quadrats

C Place quadrats where limpets occurred

D Repeat the quadrat sampling several times.

25. Which of the following is a source of variation in a species of mammal?

A Isolation

B Natural selection

C Mutation

D Adaptation.

N5

N5 Biology

Practice Papers for SQA Exams

Practice Paper A
Section 2

Fill in these boxes and read what is printed below.

Full name of centre

Town

Forename(s)

Surname

Try to answer ALL of the questions in the time allowed.

You have 2 hours and 30 minutes to complete this paper.

Write your answers in the spaces provided, including all of your working.

Leckie × Leckie
Scotland's leading educational publishers

MARKS Do not write in this margin

SECTION 2 – 75 marks
Attempt ALL questions

1. The diagram represents molecules present in a magnified fragment of cell membrane.

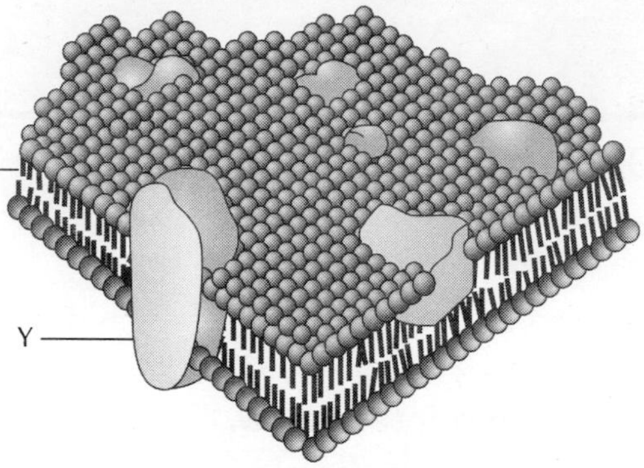

(a) Name molecules X and Y.

X ________________________________

Y ________________________________ **2**

(b) Complete the following sentences by underlining the correct options in each choice bracket.

The cell membrane is {selectively / fully} permeable and transports water in and out of the cell by osmosis.

Osmosis occurs {down / against} the concentration gradient and

and {requires / does not require} energy. **2**

(c) The diagram below shows a cell from a piece of plant tissue.

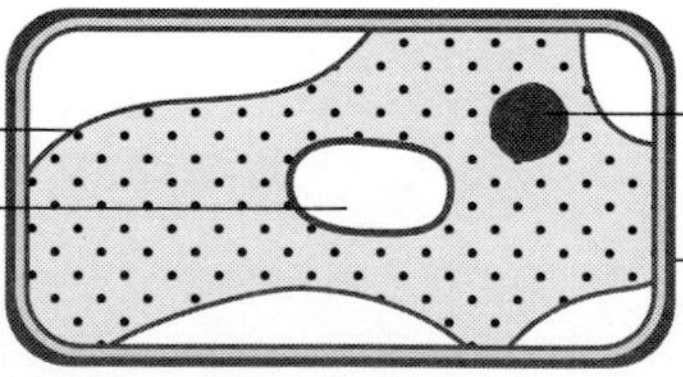

(i) Describe how a piece of plant tissue could be treated so that its cells appeared as shown in the diagram.

____

________________________________ **1**

(ii) Give the term applied to cells that appear as shown in the diagram.

________________________________ **1**

Total marks 6

MARKS
Do not write in this margin

2. The diagrams below represent stages in a synthesis (building up) reaction catalysed by a human enzyme molecule at 37 °C.

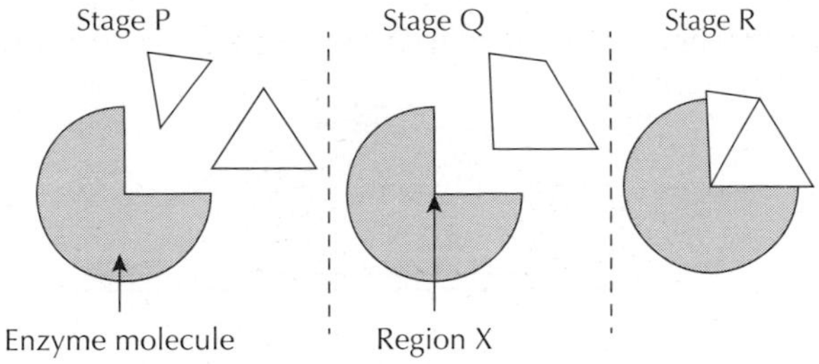

(a) Complete the flow chart below by adding letters to show the correct order of these stages as they would occur during the reaction.

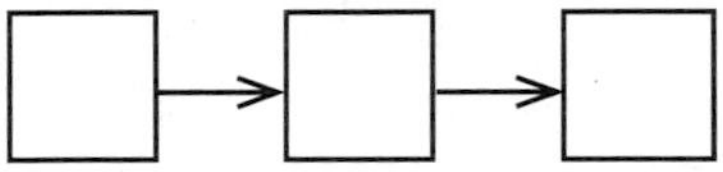

1

(b) Identify the part of the enzyme molecule labelled Region X in the diagram.

________________________________ 1

(c) Give the term which would be used to describe Stage R.

________________________________ 1

(d) Explain why the cellular reaction above would **not** occur if the temperature were increased to 60 °C.

____

____ 2

(e) Apart from temperature, give **one** other factor which could affect the rate of an enzyme- catalysed reaction.

________________________________ 1

Total marks 6

MARKS Do not write in this margin

3. An investigation was carried out on the effect of temperature on the rate of fermentation in yeast.

Apparatus as shown in the diagram below was set up, and the number of bubbles of gas produced by the yeast per minute was counted at various temperatures, as shown in the table.

Apparatus

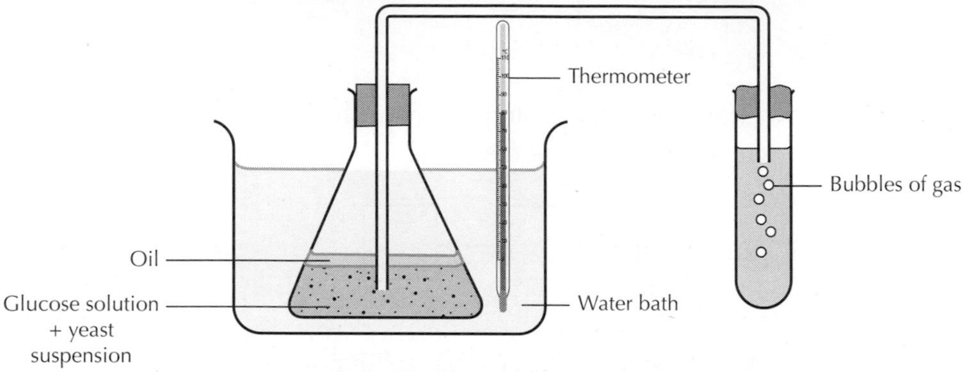

Temperature (°C)	***Bubbles of gas produced per minute***
10	30
15	50
20	80
25	110
30	120

MARKS
Do not write in this margin

(a) On the grid provided below, complete the line graph to show temperature against number of bubbles of gas produced per minute.

(A spare grid, if required, can be found at the end of the practice paper.)

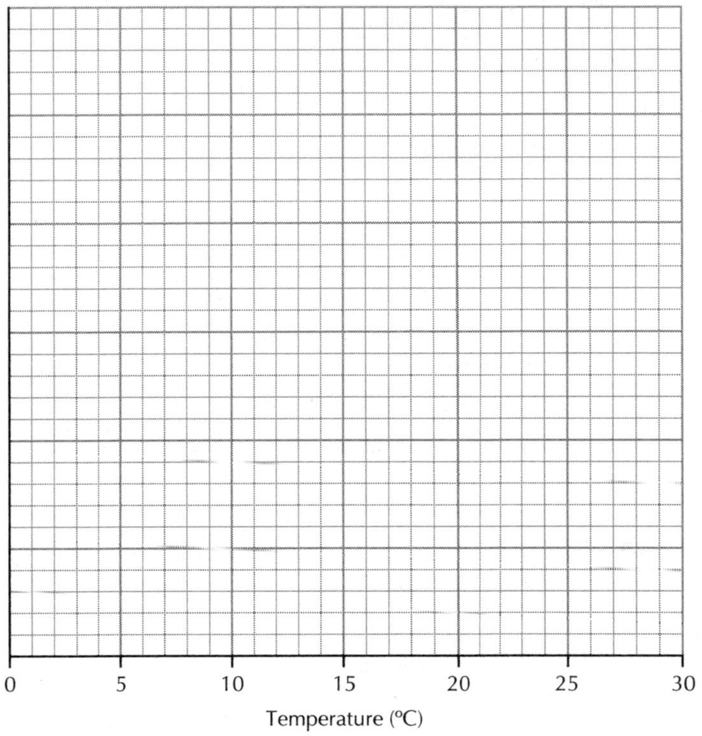

2

(b) Identify the gas produced during fermentation.

____ 1

(c) Suggest how the investigation could be improved to give more accurate results.

____

____ 1

(d) Predict how the results would be different if the investigation were repeated at 5 °C. Explain your answer.

Prediction ______________________________________ 1

Explanation

____

____ 1

Total marks 6

MARKS Do not write in this margin

4. In an investigation into beer production, a fermenter containing a glucose solution and a yeast culture was set up. The concentrations of glucose and ethanol were recorded over a 400- hour period and the results are shown in the graph below.

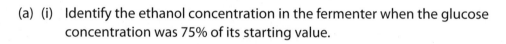

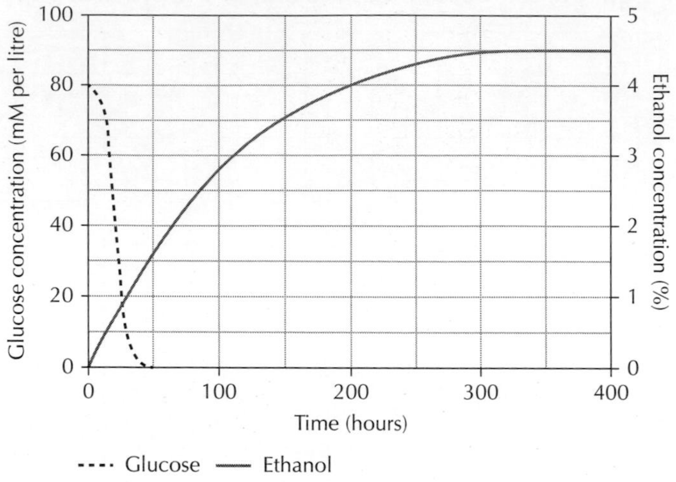

(a) (i) Identify the ethanol concentration in the fermenter when the glucose concentration was 75% of its starting value.

________________ % **1**

(ii) Give the time taken for the glucose to be completely removed from the solution.

____________ hours **1**

(iii) Calculate the average rate of ethanol production per hour over the first 200 hours of the investigation. **1**

space for calculation

____________ % ethanol per hour

(b) Describe the evidence which suggests that the yeast takes up glucose rapidly before fermenting it more slowly.

____

____ **2**

(c) Give the ethanol concentration in beer made from this fermentation.

________________ % **1**

Total marks 6

MARKS
Do not write in this margin

5. The diagram below shows a reflex arc in a human and the neurons involved.

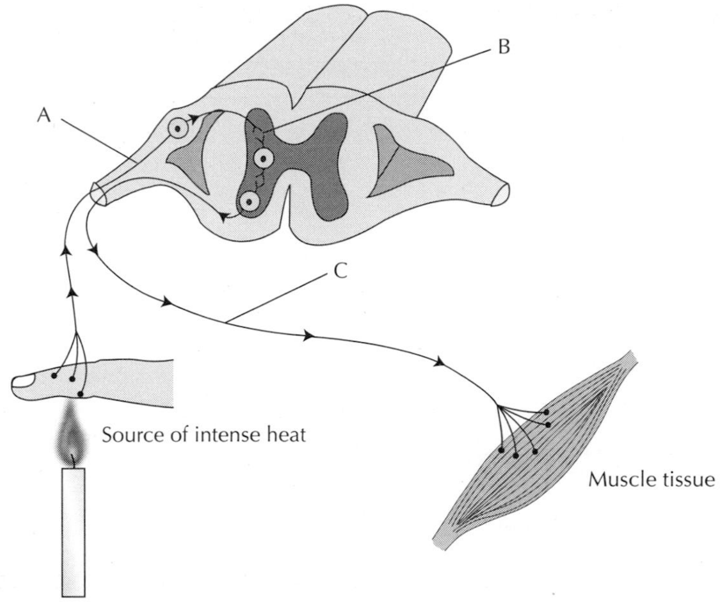

(a) Identify the type of neuron shown at A.

____________________________ **1**

(b) Name the gap at B and describe the role of chemicals that enter this gap.

Name ________________________

Role ____ **2**

(c) Explain the advantage of this reflex to the human involved.

____

____ **2**

Total marks 5

MARKS
Do not write in this margin

6. Tongue-rolling in humans is controlled by a single gene.
The dominant allele is tongue-rolling (**R**) and the recessive allele is non-rolling (**r**).

The diagram below shows the inheritance of tongue-rolling in part of a family.

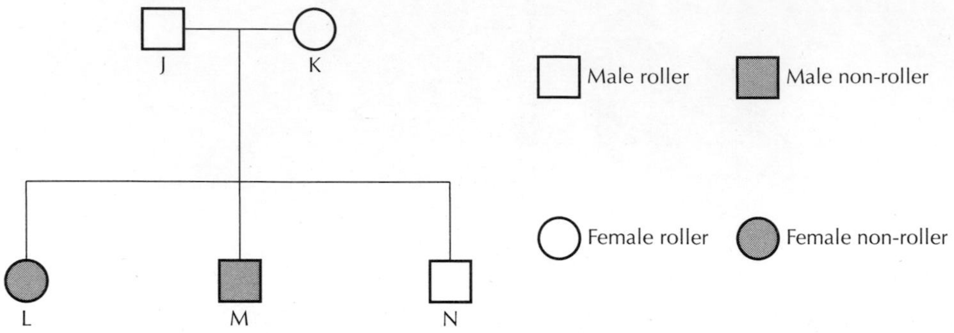

(a) Give the genotypes of the following individuals.

J ________ K ________ L ________ **2**

(b) Give the term used to describe the genotype of individuals such as L and M.

____________________ **1**

(c) Explain why, from the information given, the genotype of individual N cannot be known for certain.

____

____________________ **2**

Total marks 5

MARKS
Do not write in this margin

7. The diagram below shows cells from tissues involved in the transport of substances in a plant stem.

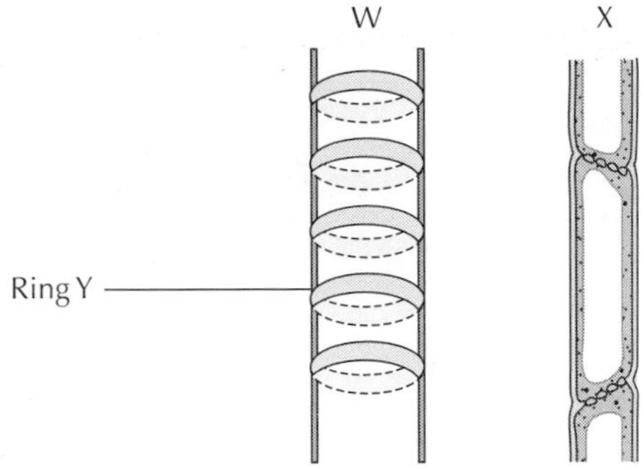

(a) Complete the table below to name tissues W and X and give **one** substance transported by each. **3**

Tissue	***Name***	***Substance transported***
W		
X		

(b) Name the substance of which ring Y is composed.

____ **1**

Total marks 4

MARKS
Do not write in this margin

8. (a) The diagram below shows cells associated with a gas exchange surface in human lung tissue following inhalation of a breath of air.

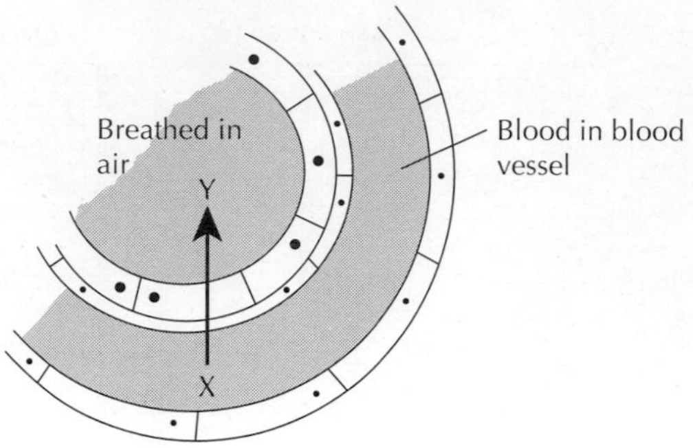

(i) Name the gas produced during cell respiration which diffuses from X to Y.

______________________________ **1**

(ii) Name the structure which contains the breathed in air shown in the diagram.

______________________________ **1**

(iii) (1) Give **one** feature of the gas exchange surface, **not shown in the diagram**, which increases its efficiency of absorption.

______________________________ **1**

(2) Give **one** feature of the blood vessel **shown in the diagram** which increases its efficiency at exchanging materials.

______________________________ **1**

MARKS
Do not write in this margin

(b) The chart below shows some information relating to the annual death rate of males in an area of the UK from coronary heart disease over the course of one year.

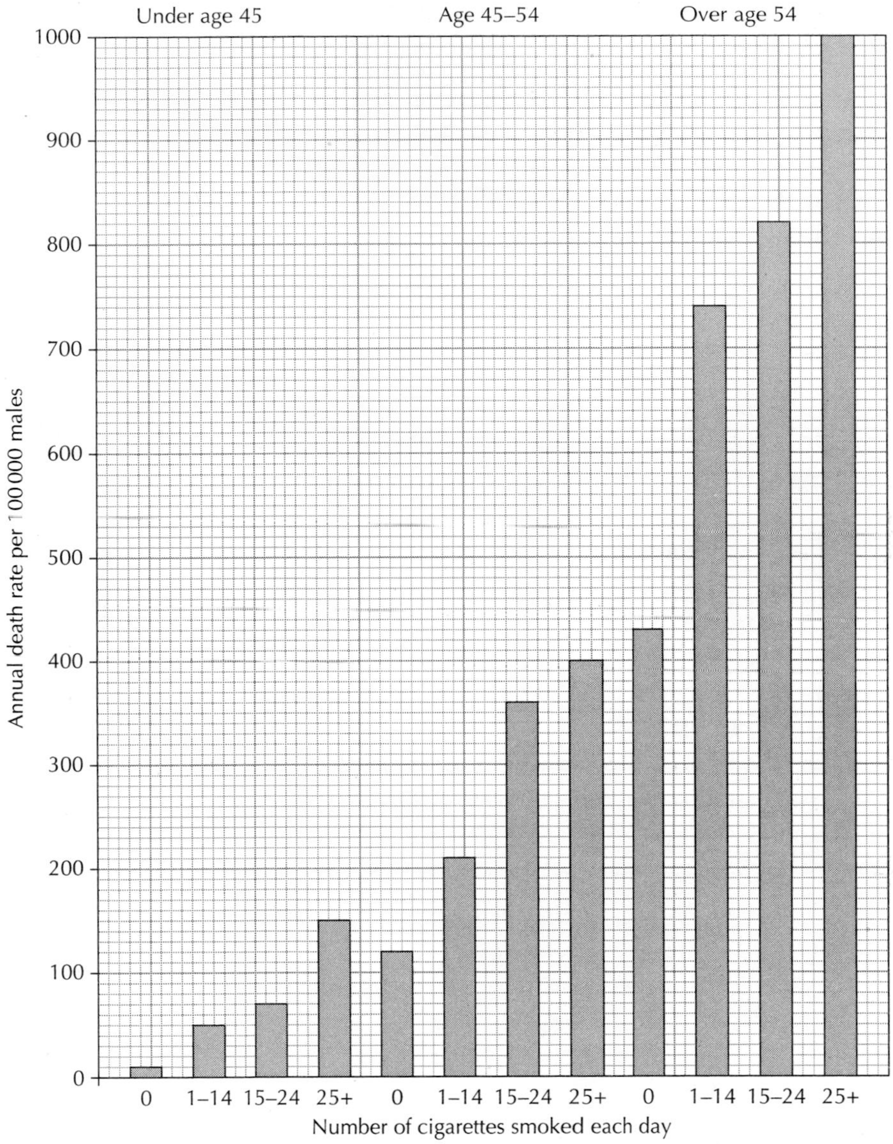

(i) From the data, identify **two** factors that affect the death rate from coronary heart disease.

1 ______________________________

2 ______________________________ **1**

(ii) Calculate the percentage increase in death rate in males aged under 45 years when the number of cigarettes smoked per day is increased from 1–14 to 25+.

Space for calculation

_______% **1**

Total marks 6

MARKS
Do not write in this margin

9. The diagram below shows cells from a sample of human blood.

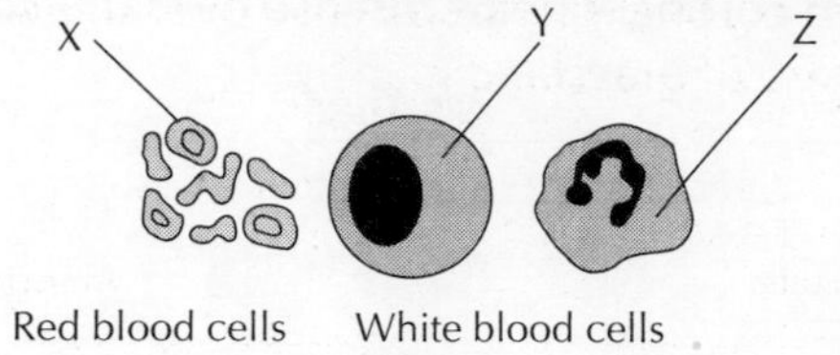

(a) Describe the shape of cell X and explain how that shape allows it to carry out its function efficiently.

Shape ____ 1

Explanation ____

____ 2

(b) Complete the table below to show the names and functions of white cells Y and Z. 2

Cell	***Name***	***Function***
Y		produces antibodies which destroy pathogens
Z	phagocyte	

Total marks 5

MARKS Do not write in this margin

10. The bar charts below show the results of an investigation carried out to compare the numbers of four different species of ground layer plants in a hectare of woodland with the numbers found in a hectare of grassland nearby.

(1 hectare = 10 000 m²)

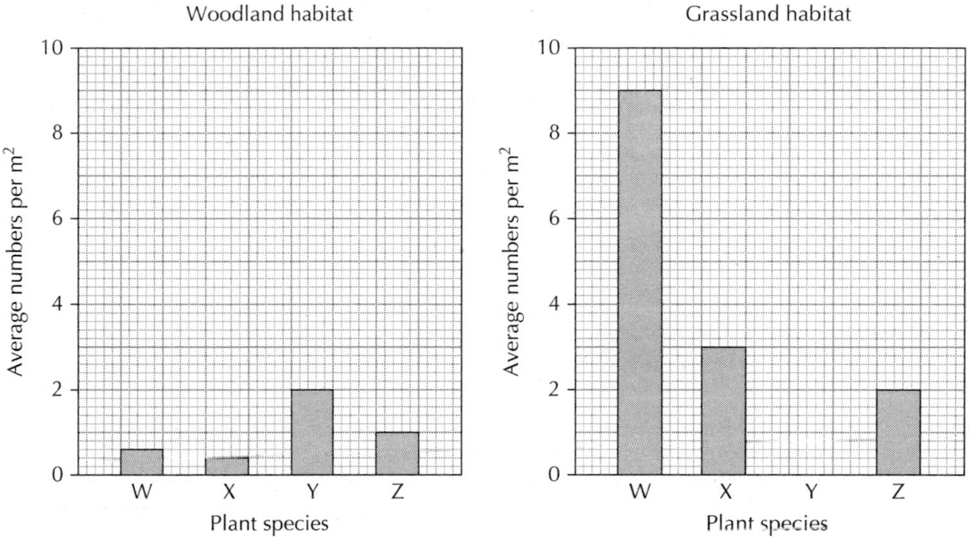

Key:	
W Perennial rye grass	X Clover
Y Wood sorrel	Z Creeping buttercup

(a) Name a technique that could be used to estimate the number of plants of each species present and describe its use.

Name ____________________

Description ____________________ **1**

(b) Calculate the simplest whole number ratio of perennial rye grass to clover in the grassland habitat.

Space for calculation

______ : ______
perennial rye grass   clover **1**

(c) Estimate the total number of wood sorrel plants that would be present in the entire hectare of woodland.

Space for calculation

__________ plants **1**

MARKS
Do not write in this margin

(d) **Choose** an abiotic factor that might be involved in the different abundance of perennial rye grass in these two habitats and explain its role.

Abiotic factor ______________________

Explanation ____

____ 2

Total marks 5

MARKS Do not write in this margin

11. The diagram below shows a slow-flowing stream passing through an area of farmland and five sampling sites along the stream. Fertiliser was applied to the land only in the area shown.

Six months after the fertiliser was applied, the water in the stream was sampled and its oxygen and bacterial contents measured. The results from the five sites are shown in the table below.

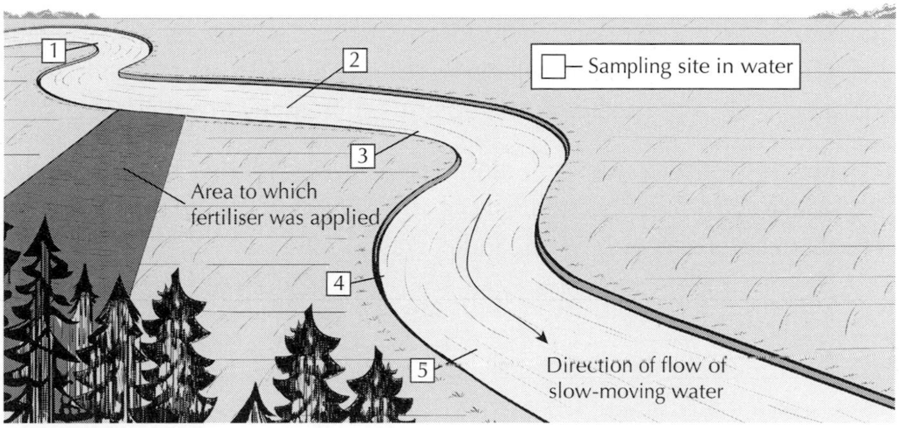

Sampling site	***Oxygen level (units)***	***Bacterial numbers***
1	140 000	low
2	400	very high
3	800	high
4	4500	high
5	16 000	low

(a) It was suggested that fertiliser from the farmland had entered the stream. Explain how this might have happened.

____ **1**

(b) Use information from the table to describe how the fertiliser from the fields might have caused the changes in oxygen levels shown in the table between Sampling site 1 and 2 over the six months.

____

____

____ **3**

(c) Identify evidence from the table which suggests that the effects of the fertiliser might not be permanent.

____ **1**

Total marks 5

MARKS Do not write in this margin

12. Read the following passage and answer the questions based on it.

Biodiversity indicators

Terrestrial breeding birds are a good indicator of overall biodiversity. Birds respond quickly to variation in habitat quality, through changes in breeding success, survival or distribution. Since most bird species are relatively easy to identify and count and are abundant and active during daytime, they are often used as indicators of biodiversity.

Terrestrial breeding birds in Scotland include familiar garden species such as blackbird and robin, woodland species such as willow warbler and goldcrest, farmland species such as linnet and goldfinch, and species of the uplands such as raven and black grouse .

The index of numbers of terrestrial breeding birds is used as an indicator of biodiversity. The index compares bird numbers against the 1994 figure which is taken as 100. In 2007 the index stood at 121.1 and in 2015 the index was 118.2.

(a) (i) Give **two** reasons why terrestrial birds are used as indicators of biodiversity.

1 ____________________

2 ____________________ **2**

(ii) Give **one** environmental factor, not mentioned in the passage, which is monitored using indicator species.

____________________ **1**

(b) Give **one** example of a bird species which breeds in farmland habitats.

____________________ **1**

(c) Calculate the difference in the terrestial bird index between the following dates:

(i) 2007 and 2015

____________________ **1**

(ii) 1994 and 2015

____________________ **1**

Total marks 6

MARKS
Do not write in this margin

13. The diagram below shows parts of two stages of photosynthesis in a green plant.

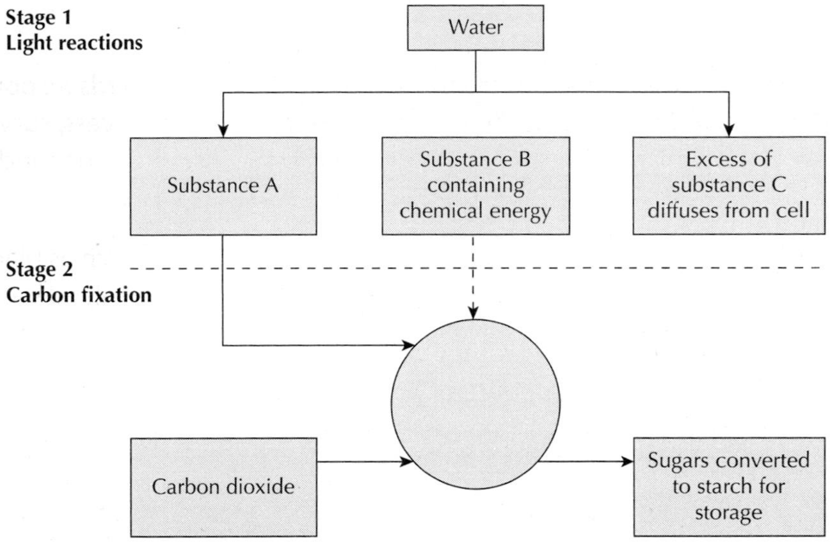

(a) Complete the table below by naming substances A, B and C produced during Stage 1.

Substance	***Name***
A	
B	
C	

2

(b) The experiment shown in the diagram below was used to investigate the requirements for photosynthesis in a green plant.

The plant was kept in darkness for 24 hours before being placed in bright light for 5 hours.

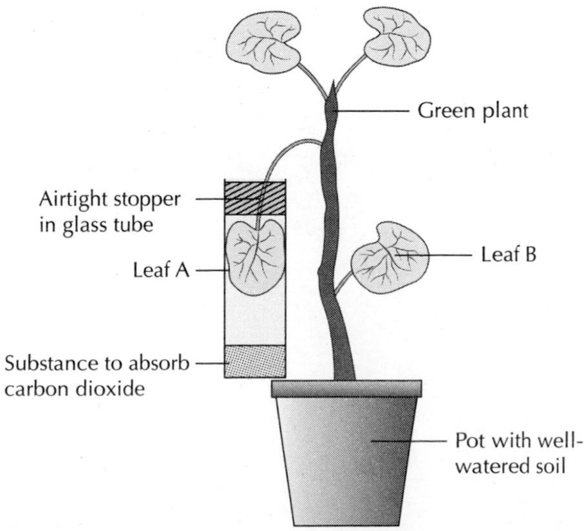

(i) After the apparatus had been in bright light for 5 hours, a test for starch was carried out on **leaf A**.

Predict whether a positive or negative starch test would be obtained and give a valid conclusion about the requirements for photosynthesis that can be drawn from it.

Result ______________________ 1

Conclusion ______________________________________

______________________________________ 1

(ii) Describe how **leaf B** would be treated so that it could act as a control in this experiment.

____

______________________________________ 1

(iii) Describe how the apparatus could be altered to show that light is needed for photosynthesis.

____

______________________________________ 1

____

Total marks **6**

MARKS Do not write in this margin

14. The table and diagrams below give information about the beaks of two species of finch and a description of the habitats they occupy on the Galapagos Islands.

Size and shape of beak	***Description of habitat***
wide, deep and blunt	woodland with flowering shrubs providing large seeds and nuts
long, narrow and pointed	woodland with rotting logs providing food for insects

Finch species P

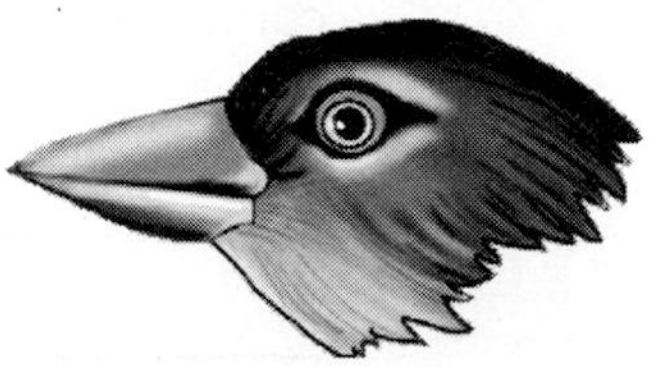

Finch species Q

(a) Identify the finch species that eats large seeds and give a reason for your choice.

Species ___________

Reason ______________________________

______________________________ **1**

(b) Suggest **two** ways in which competition between the two species is reduced.

1 ______________________________

2 ______________________________ **2**

(c) These two species may have arisen by evolution from a common ancestor. The processes below are involved in the formation of new species.

P Mutation

Q Natural selection

R Isolation

Complete the flow chart below by adding the letters to show the order in which these processes would have occurred to produce the two species of finch.

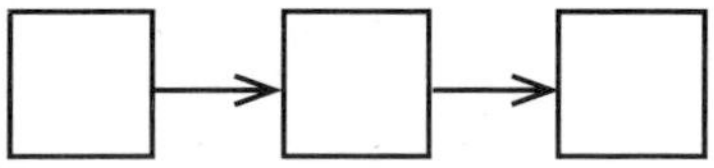

1

Total marks 4

[END OF QUESTION PAPER]

MARKS
Do not write in this margin

ADDITIONAL GRAPH PAPER

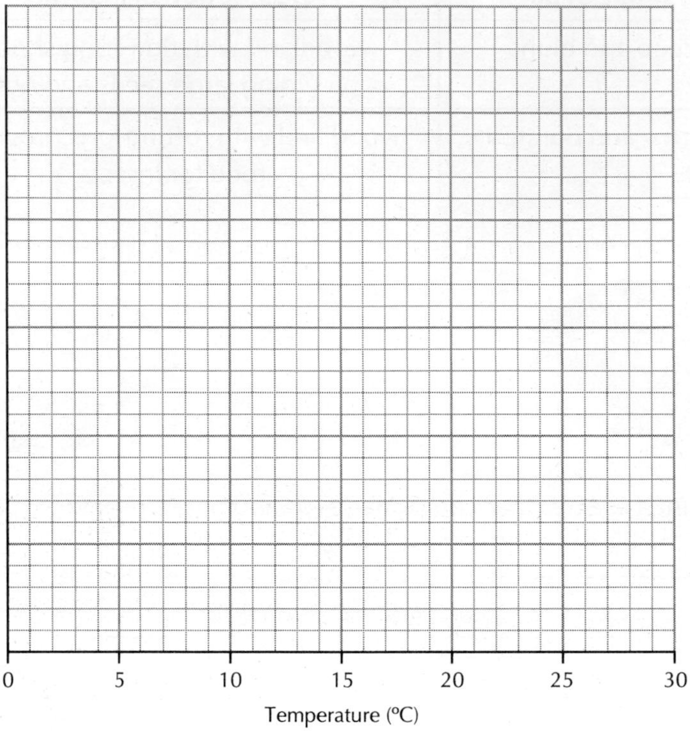

Practice Paper B

SECTION 1 ANSWER GRID

Mark the correct answer as shown ✓

	A	B	C	D
1	○	○	○	○
2	○	○	○	○
3	○	○	○	○
4	○	○	○	○
5	○	○	○	○
6	○	○	○	○
7	○	○	○	○
8	○	○	○	○
9	○	○	○	○
10	○	○	○	○
11	○	○	○	○
12	○	○	○	○
13	○	○	○	○
14	○	○	○	○
15	○	○	○	○
16	○	○	○	○
17	○	○	○	○
18	○	○	○	○
19	○	○	○	○
20	○	○	○	○
21	○	○	○	○
22	○	○	○	○
23	○	○	○	○
24	○	○	○	○
25	○	○	○	○

N5

N5 Biology

Practice Papers for SQA Exams

Practice Paper B
Section 1

Fill in these boxes and read what is printed below.

Full name of centre	Town

Forename(s)	Surname

Try to answer ALL of the questions in the time allowed.

You have 2 hours and 30 minutes to complete this paper.

Write your answers in the spaces provided, including all of your working.

Leckie × Leckie
Scotland's leading educational publishers

SECTION 1 – 25 marks
Attempt ALL questions

1. The diagram below shows some structures present in a fungal cell.

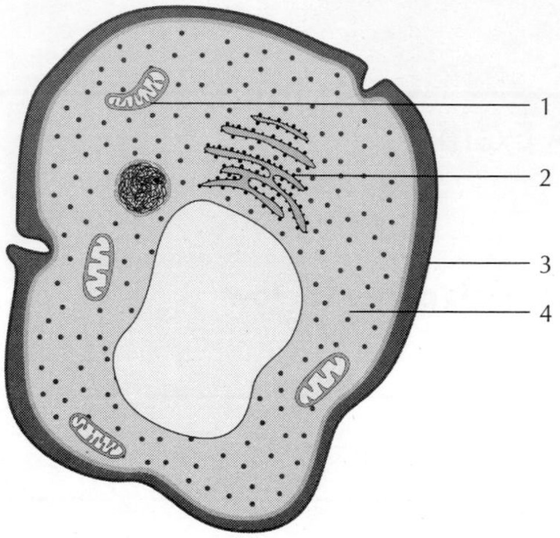

Which line in the table below identifies correctly the site of aerobic respiration and the structure that provides support for the cell?

	Site of aerobic respiration	*Provides support for cell*
A	1	4
B	2	3
C	2	4
D	1	3

2. The diagram below shows cells in a piece of onion epidermal tissue as seen under a microscope.

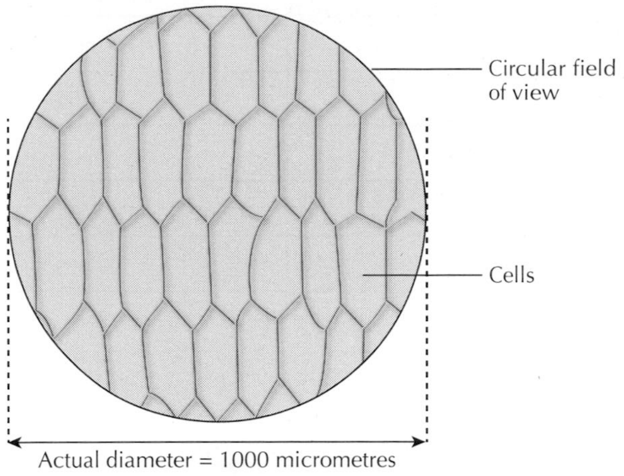

The best estimate of the average **width** of the cells shown is

A 10 micrometres

B 25 micrometres

C 100 micrometres

D 250 micrometres

3. **Questions 3 and 4 refer to the diagram below which shows molecules present in a section of highly magnified cell membrane.**

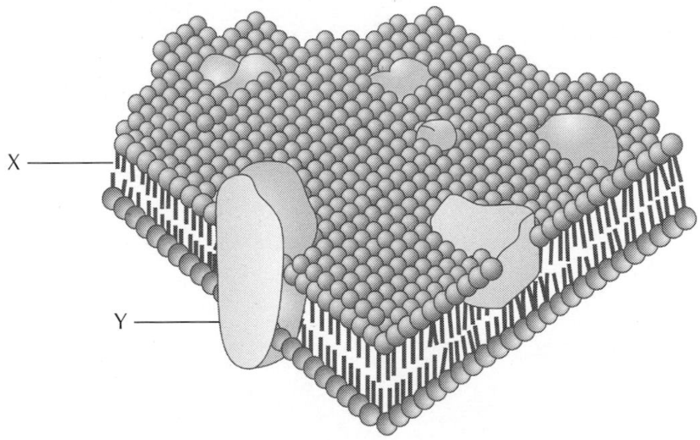

Which line in the table below correctly identifies these molecules?

	Molecule X	*Molecule Y*
A	protein	phospholipid
B	phospholipid	cellulose
C	protein	cellulose
D	phospholipid	protein

4. Which line in the table shows requirements for active transport?

	Energy required	***Membrane proteins required***
A	Yes	No
B	No	No
C	Yes	Yes
D	No	Yes

5. Which line in the table below shows correctly the terms that apply to the descriptions of enzyme action given?

	Description of enzyme action	
	best conditions for enzyme action	***effect of reaction on enzyme molecules***
A	optimum	denatured
B	specific	unchanged
C	optimum	unchanged
D	specific	denatured

6. The diagram below shows stages in an enzyme-catalysed synthesis reaction.

$P + Q + R \rightarrow S \rightarrow T + R$

Which of the following represents the enzyme-substrate complex?

A $P + Q + R$

B S

C $P + Q$

D $T + R$

7. The following are stages in the genetic engineering of bacteria.

1 Insert plasmid into host cell

2 Extract required gene from chromosome

3 Insert required gene into plasmid

4 Remove plasmid from host cell.

Which is the correct sequence of stages that would be carried out during the process of genetic modification of the bacteria?

A 2→4→1→3

B 2→4→3→1

C 3→4→1→2

D 3→1→4→2

8. The diagram below shows respiratory pathways in a mammalian muscle cell.

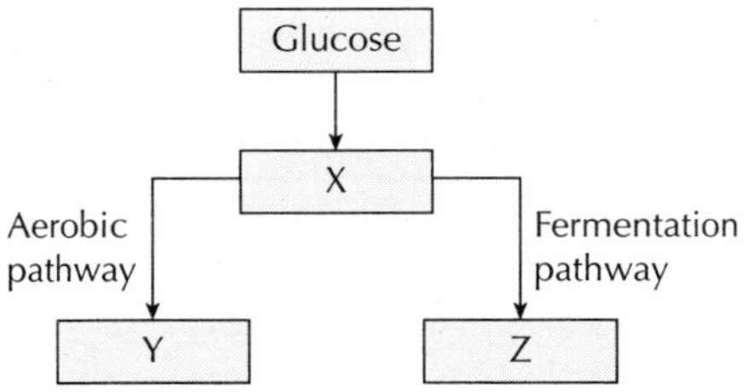

Which line in the table below identifies correctly the substances in boxes X, Y and Z?

	X	*Y*	*Z*
A	pyruvate	carbon dioxide and water	lactate
B	pyruvate	lactate	carbon dioxide and water
C	lactate	pyruvate	carbon dioxide and water
D	carbon dioxide and water	pyruvate	lactate

9. The respirometer shown below was used in an investigation of respiration in yeast.

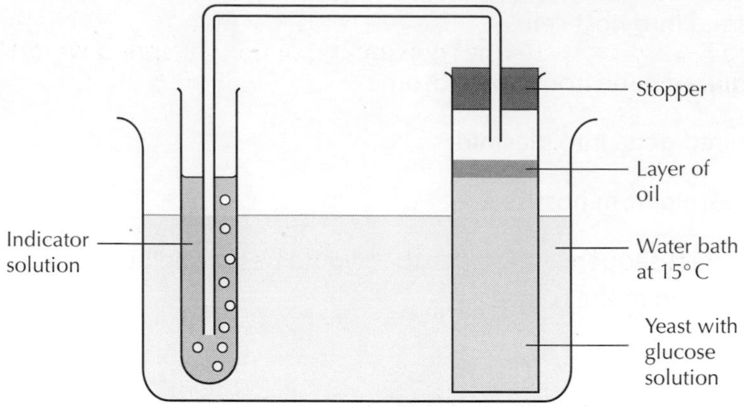

Which change to the apparatus would cause a **decrease** in the respiration rate of yeast?

A Leaving out the oil layer

B Diluting the glucose solution

C Increasing the water bath temperature to 20 °C

D Using cotton wool instead of a rubber stopper.

10. Which of the following statements about stem cells is **not** correct.

Stem cells

A are specialised cells

B divide by mitosis to self-renew

C are involved in growth and repair

D can become different types of cell.

11. A homozygous black-coated male mouse was crossed with a homozygous brown-coated female.

All the F_1 mice were black.

The F_1 mice were allowed to mate, and the F_2 generation contained both black and brown mice.

What evidence is there that the allele for black coat is dominant to the allele for brown coat?

A Only one of the original parents was black

B The original male parent was black

C All of the F_1 mice were black

D Some of the F_2 mice were black.

12. In a breeding experiment with *Drosophila*, homozygous normal winged flies were crossed with homozygous vestigial winged flies. All of the F_1 were normal winged.

If flies from the F_1 were crossed, what percentage of their offspring would be expected to have normal wings?

A 25%

B 50%

C 75%

D 100%

13. The diagram below shows a vertical section through a flower.

Which part produces male gametes?

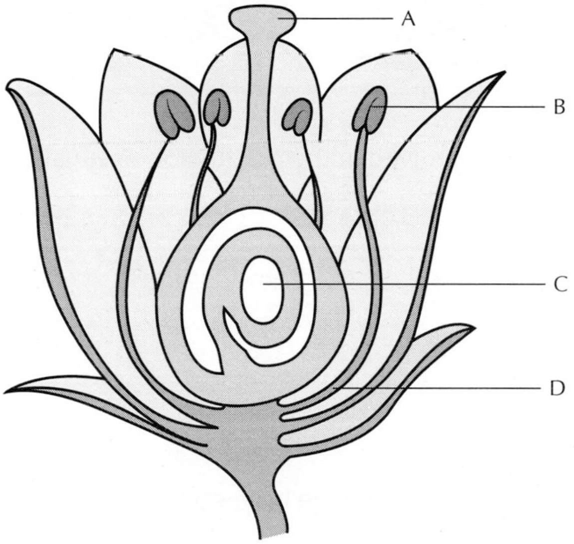

14. The diagram below shows a section through a green leaf.

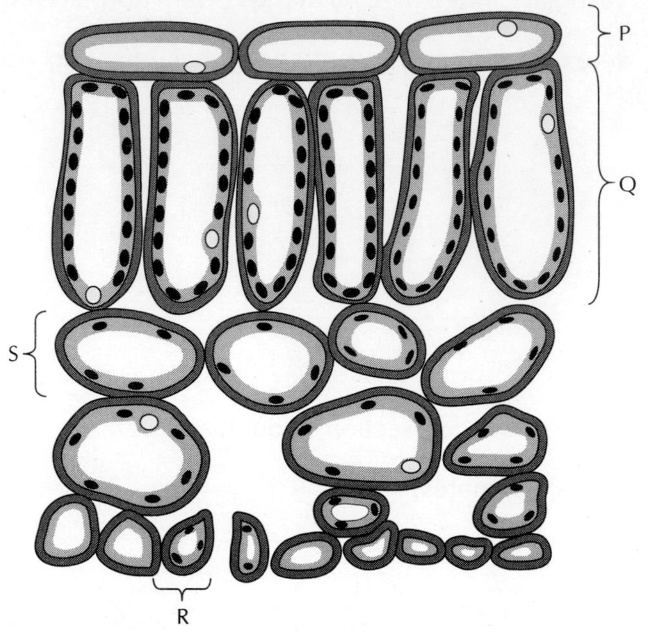

Which part of the leaf is **not** involved in the production of sugar by photosynthesis?

A P

B Q

C R

D S

15. The plant tissue that carries sugar from the leaves to the roots is the

A mesophyll

B xylem

C phloem

D epidermis.

16. The diagram below shows a single villus from the small intestine of a mammal.

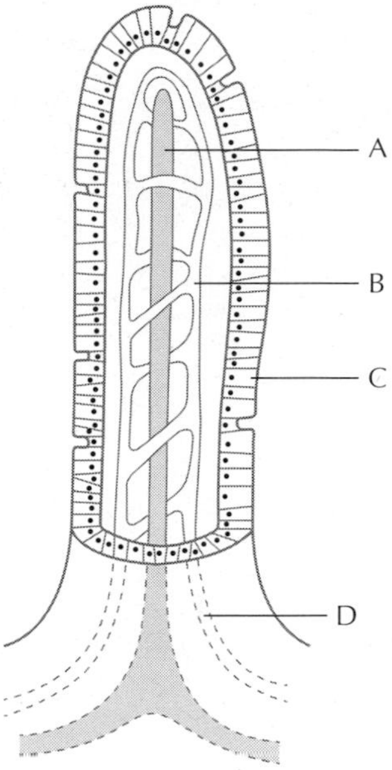

Which part is the lacteal?

17. Which line in the table below identifies correctly examples of biotic and abiotic factors affecting biodiversity?

	Biotic	***Abiotic***
A	temperature	grazing
B	pH	temperature
C	grazing	predation
D	predation	pH

18. The diagram below represents a pyramid of numbers.

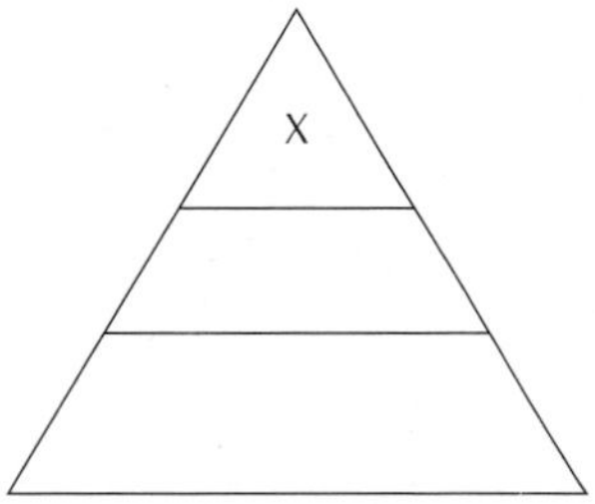

The block at X best represents the total numbers of

A producers

B herbivores

C predators

D prey.

19. Which labelled organism in the food web shown below has the **least** number of interspecific competitors for each food source?

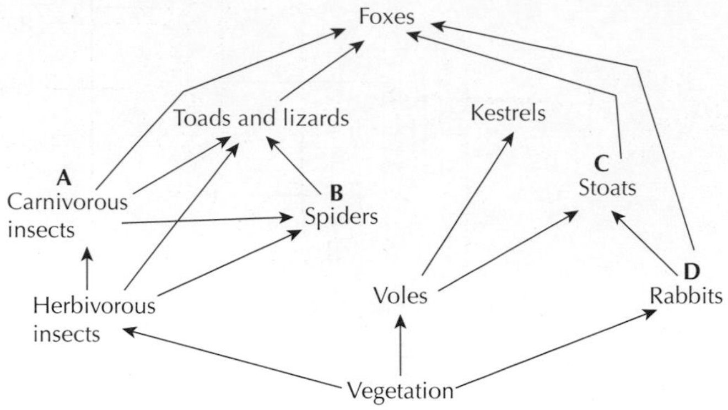

20. Various aspects of a river were sampled at five points. The results are shown in the table below.

Aspect sampled	***Sampling points***				
	1	2	3	4	5
Mayfly nymph number	89	15	0	0	0
Midge larvae numbers	0	1	2	175	24
Oxygen concentration (% of maximum)	85	85	75	30	63
pH level	5.5	6.0	6.4	7.3	8.0

Based on the results in the table, which of the following conclusions is valid?

A High oxygen concentration limits the numbers of midge larvae

B pH level is proportional to oxygen concentration

C Midge larvae do not survive in water with low oxygen concentration

D Mayfly numbers depend on oxygen concentration alone.

21. The apparatus below was set up to investigate photosynthesis in an aquatic plant.

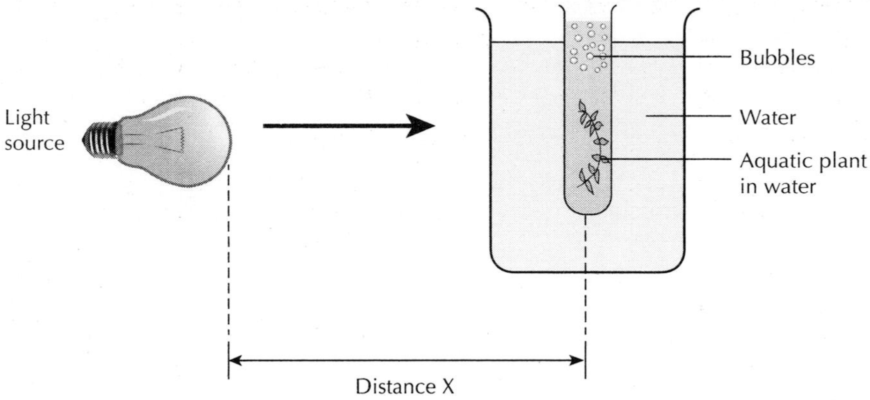

The list below shows variables related to photosynthesis that could be measured.

1 Light intensity

2 Rate of carbon fixation

3 Rate of bubble production

If distance X was increased, which variable(s) on the list would **decrease**?

A 1 only

B 2 only

C 1 and 3 only

D 1, 2 and 3

22. The role of chlorophyll in photosynthesis is to trap

A light energy for ATP production

B light energy for carbon dioxide absorption

C chemical energy for carbon dioxide absorption

D chemical energy for ATP production.

23. The following stages are involved in speciation.

1 Natural selection

2 Isolation

3 Mutation

In which order do these occur?

A 2→3→1

B 1→2→3

C 2→1→3

D 3→2→1

24. The graph below shows the increase in the human population between the years 1400 and 2000.

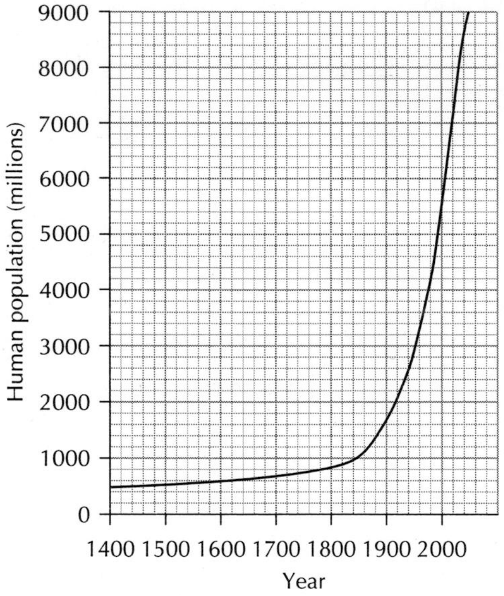

What was the percentage increase in the population between 1850 and 1950?

A 60%

B 200%

C 250%

D 50%

25. In an investigation into intraspecific competition, a batch of cress seeds of the same variety were planted out into three containers, as shown below. The containers were well watered then placed together in a bright evenly-lit room.

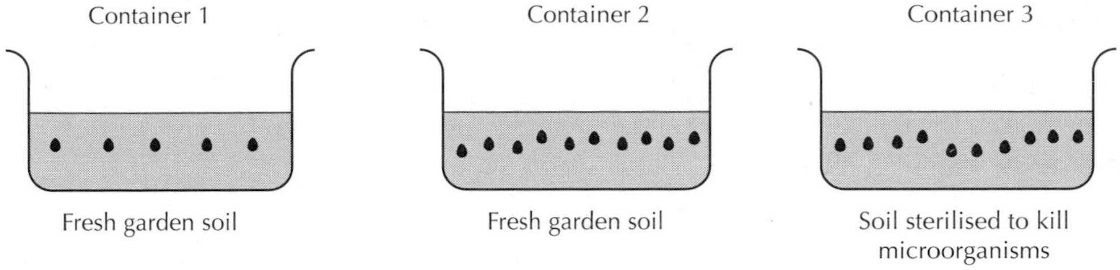

The diagrams below show the appearance of the containers after three days.

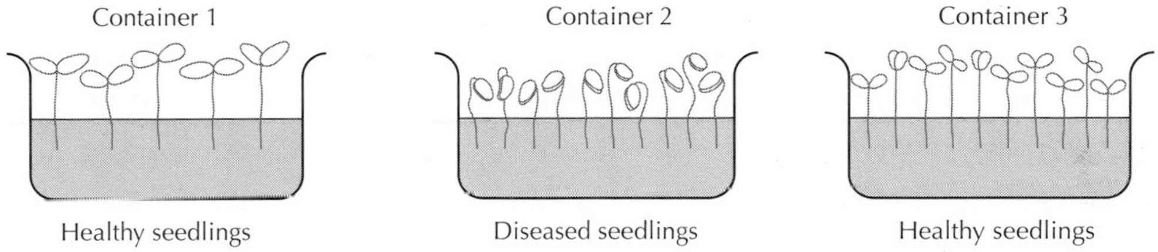

Which line in the table below correctly identifies the factor(s) involved in the diseased state of the seedlings in Container 2?

	Factors		
	sowing density	***microorganisms in soil***	***light intensity***
A	✔	✔	✔
B	✔	✕	✕
C	✕	✔	✕
D	✔	✔	✕

Key

✔ Factor involved
✕ Factor not involved

N5

N5 Biology

Practice Papers for SQA Exams

Practice Paper B
Section 2

Fill in these boxes and read what is printed below.

Full name of centre

Town

Forename(s)

Surname

Try to answer ALL of the questions in the time allowed.

You have 2 hours and 30 minutes and thirty minutes to complete this paper.

Write your answers in the spaces provided, including all of your working.

Leckie × Leckie
Scotland's leading educational publishers

MARKS
Do not write in this margin

SECTION 2 – 75 marks
Attempt ALL questions

1. Thin pieces of onion epidermis were immersed in solutions, as shown in the diagram below, and left for one hour.

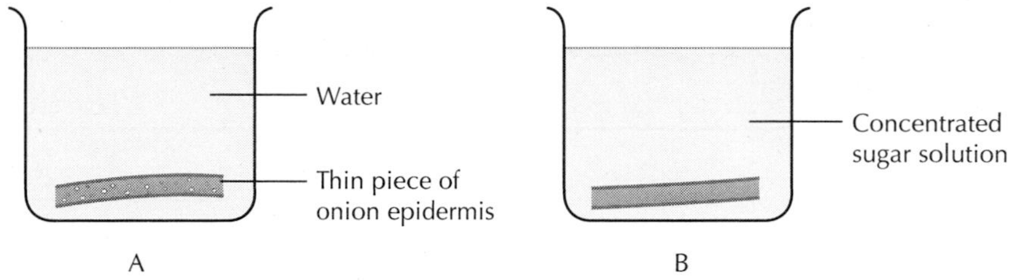

(a) The diagram below shows the appearance of an onion cell from dish A after one hour.

Complete the diagram to show the predicted appearance of a cell from dish B after this time.

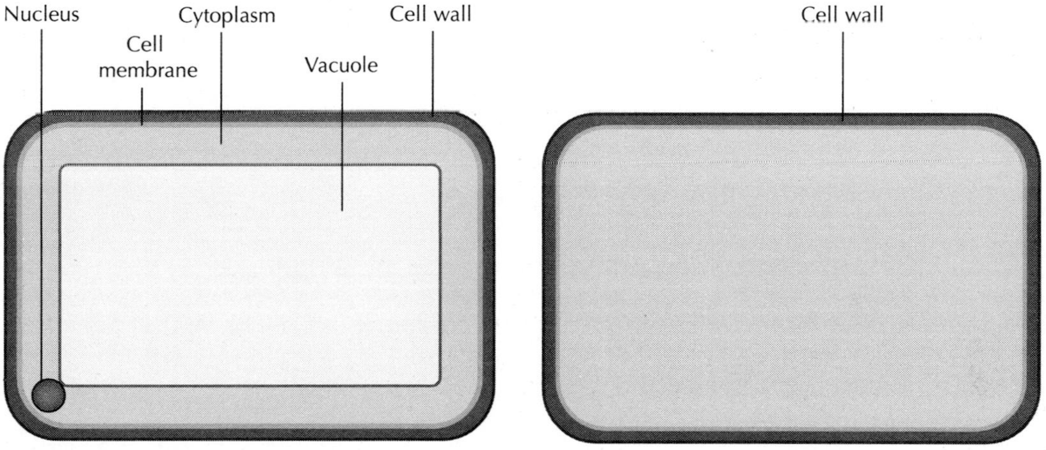

Cell from dish A   Cell from dish B   **2**

(b) Give the term used to describe the state of a cell, such as that from dish A, which has been immersed in pure water for one hour.

____ **1**

MARKS
Do not write in this margin

(c) (i) Name the process that has led to the different appearances of the onion cells in dishes A and B.

_______________________________________ **1**

(ii) The process responsible for these changes is described as being passive. Give the meaning of the term passive in this example.

____

_______________________________________ **1**

(d) Complete the following sentence by <u>underlining</u> the correct option in each choice bracket.

Onion epidermis is a(n) { organ / tissue } which is made up of cells carrying out a { similar / different } function. **1**

Total marks 6

MARKS Do not write in this margin

2. The diagram below represents cells from a region of cell division in a young plant root.

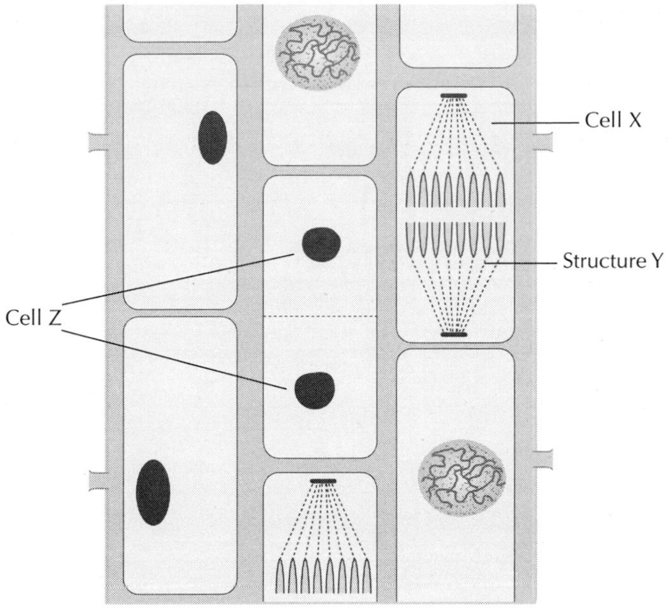

(a) Describe the stage of mitosis shown in cell **X**.

____

____ **1**

(b) Name the structure labelled **Y**.

______________________________ **1**

(c) Complete the sentences below by underlining the correct option in each of the choice brackets.

The two nuclei in cell **Z** are genetically {different / identical} to each other, and each has the {haploid / diploid} number of chromosomes. The two cells that are forming will be {specialised / unspecialised}. **2**

Total marks 4

3. The diagram below shows a stage of protein synthesis in which messenger RNA (mRNA) is formed in the nucleus of a cell.

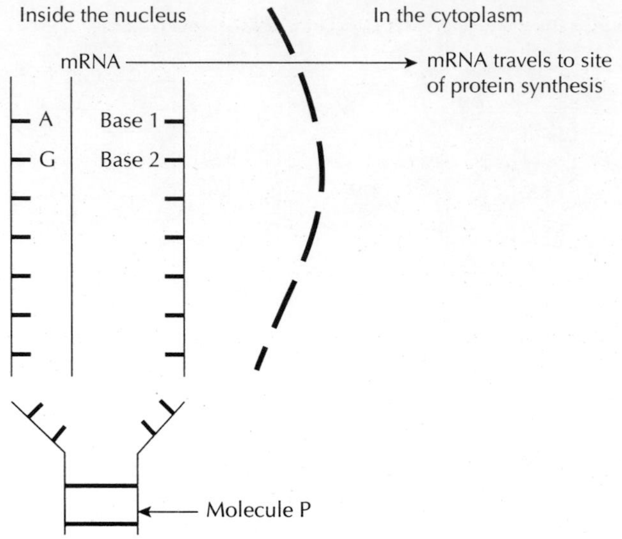

(a) Name molecule P.

_______________ 1

(b) Identify the bases 1, 2.

Base 1 _______________

Base 2 _______________ 2

(c) Once complete, the mRNA molecule leaves the nucleus and enters the cytoplasm.

Identify the cell structures to which the mRNA travels and where protein synthesis takes place.

_______________ 1

(d) Describe the feature of the mRNA molecule that ensures that the correct protein is synthesised.

____

_______________ 1

(e) Give **one** function of proteins in cells.

_______________ 1

Total marks 6

MARKS Do not write in this margin

4. An experiment was carried out to investigate the effect of temperature on the digestion of lipid (fat) by the enzyme lipase.

The experiment was set up as shown in the diagram below.

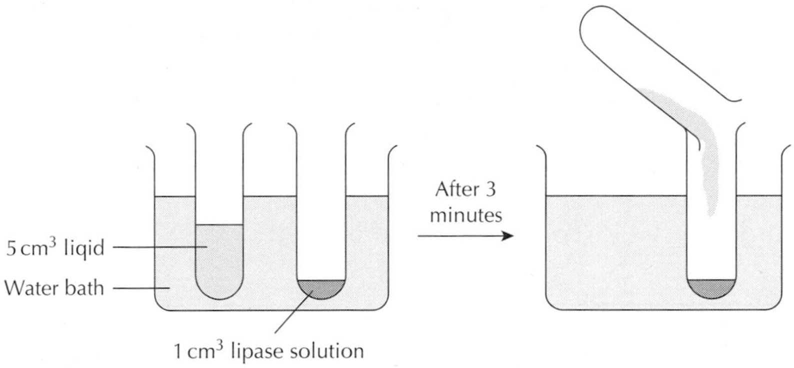

The time taken for the lipid to be fully digested was recorded and the results are shown in the table below. The experiment was carried out at various temperatures and the results are shown in the table below.

Temperature (°C)	***Average time taken until no lipid remained (minutes)***
5	40
20	20
40	5
50	30
90	lipid remained undigested after 120 minutes

(a) Explain why the test tubes of lipid and lipase solution were kept separately in the water bath for three minutes at each temperature before mixing. **1**

____

(b) Give **one** variable that should have been controlled to allow a valid conclusion to be made. **1**

____

(c) Describe the effect of increasing temperature on the activity of lipase. **1**

____

(d) Suggest **one** improvement that could increase the reliability of the results. **1**

____

Total marks 4

MARKS Do not write in this margin

5. The structure of a sperm cell is shown in the diagram below.

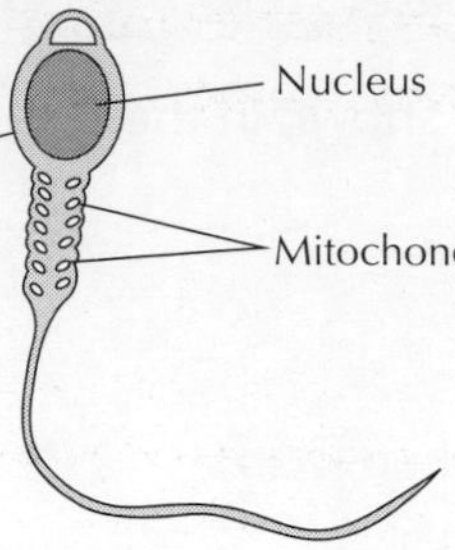

(a) Name the organs in which sperm cells are produced.

____ **1**

(b) Explain why there are a large number of mitochondria in the sperm cell.

____ **1**

(c) A sperm cell is haploid.

Explain the meaning of this statement in terms of the chromosome complement.

____

____ **1**

(d) Describe what happens during fertilisation.

____

____ **1**

Total marks 4

MARKS
Do not write in this margin

6. In an investigation of fermentation, 20 cm^3 of a yeast suspension was added to 50 cm^3 of grape juice and the carbon dioxide gas produced was collected and measured, as shown in the respirometer below.

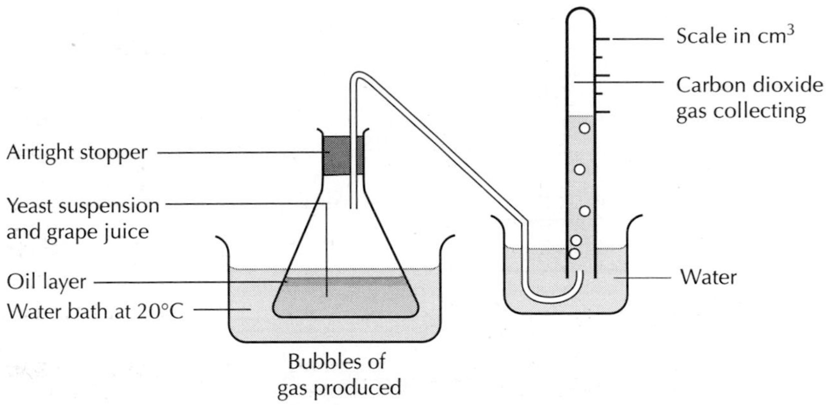

The rate of fermentation was calculated every 2 days for 10 days. The results are shown in the table below.

Day	***Rate of fermentation* (cm^3 carbon dioxide produced per hour)**
0	0
2	15
4	25
6	30
8	12
10	2

(a) On the grid below plot a line graph to show the rate of fermentation against time.

(A spare grid, if required, can be found at the end of the practice paper.)

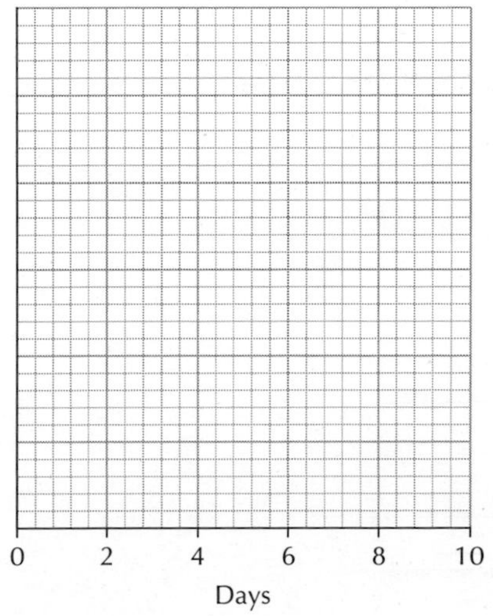

2

MARKS
Do not write in this margin

(b) Calculate the simplest whole number ratio of volume of carbon dioxide produced per hour after 2 days to that produced after 8 days.
Space for calculation

_____ : _____
2 days 8 days **1**

(c) Suggest a reason for the reduction in rate of fermentation after day 6.

____

____ **1**

(d) Suggest an improvement to the method described that would allow the investigation to be repeated more accurately.

____

____ **1**

(e) The list shows various factors that could affect the rate of respiration.

temperature   concentration of grape juice   concentration of yeast suspension

Choose a factor and describe how the apparatus could be used to investigate its effect on the rate of fermentation.

Factor chosen ____________________

Description ______________________________________

____ **1**

Total marks 6

7. The diagram below shows the ends of two neurons J and M, the gaps between them and an interneuron K within the spinal cord in the central nervous system of a mammal.

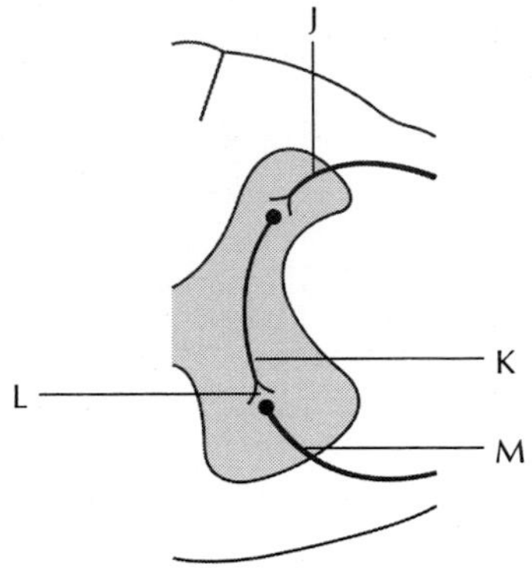

(a) Describe how a nervous message is passed along a neuron such as J.

______________________________ **1**

(b) Describe how a nervous message arriving at the end of neuron K is able to cross the gap L.

____

______________________________ **1**

(c) Name gap L.

______________________________ **1**

(d) Give **one** feature of a reflex action and explain the advantage it provides for mammals.

Feature ______________________________ **1**

Advantage ______________________________ **1**

Total marks 5

MARKS Do not write in this margin

8. The bar chart below shows variation in the length of seeds harvested from a broad bean plant.

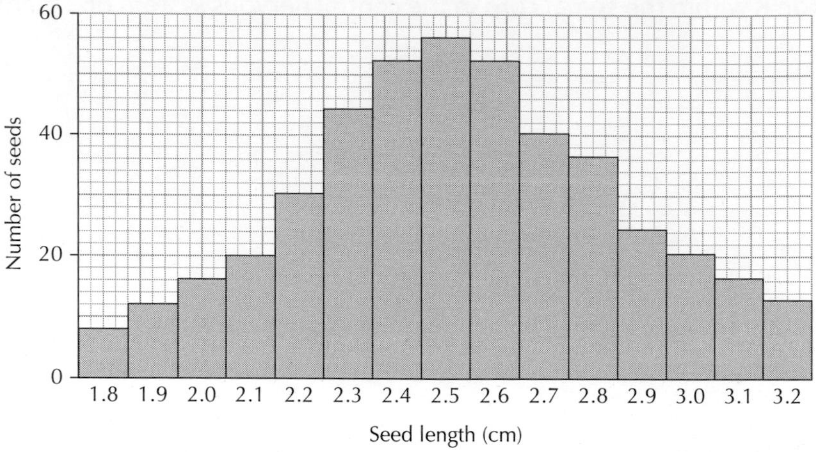

(a) Calculate the difference between the shortest and longest seeds in the sample.

Space for calculation

______ cm **1**

(b) Give evidence to support the statement that the seed length shows continuous variation.

______

______ **1**

(c) Give **one** example of a characteristic from a **named** animal or plant species that shows discrete variation.

Named species ______

Characteristic ______ **1**

Total marks 3

MARKS
Do not write in this margin

9. The diagram below shows the heart and an outline of the circulatory system of a human.

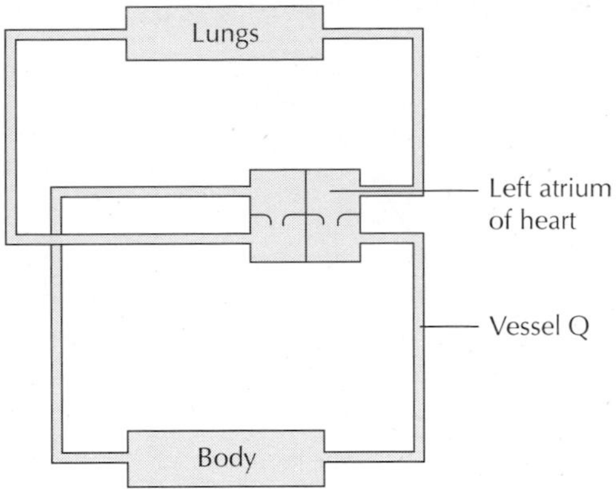

(a) **On the diagram**:

(i) Use the letter P to label the pulmonary artery. **1**

(ii) Draw an arrow on vessel Q to show the direction of blood flow. **1**

(b) Name the structures found in the heart and veins that prevent the backflow of blood.

______________________________ **1**

(c) Describe the structure of red blood cells and explain how they are adapted to take up and transport oxygen.

____

____

____

____

____ **2**

Total marks 5

MARKS
Do not write in this margin

10. The graph below shows the average transpiration rate of barley plants in an open field over a 24-hour period during summer in Scotland.

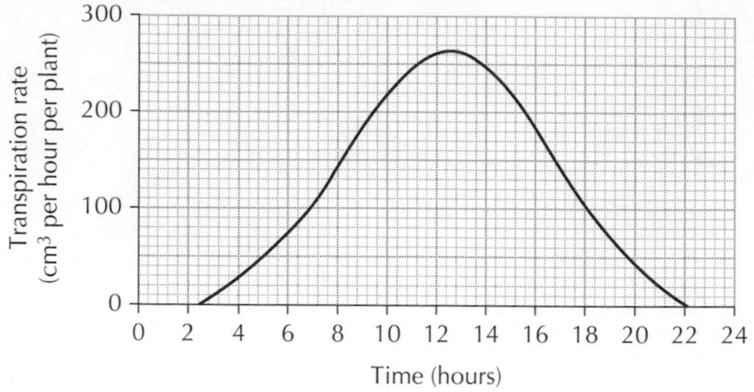

(a) Give the period during which the average transpiration rate is greater than 100 cm^3 per hour per plant.

From __________ hours until __________ hours **1**

(b) Name **two** environmental factors that might be involved in the changing rates of transpiration over the period.

1 ______________________________________

2 ______________________________________ **2**

(c) The diagram below shows cells in the lower epidermis of a barley leaf.

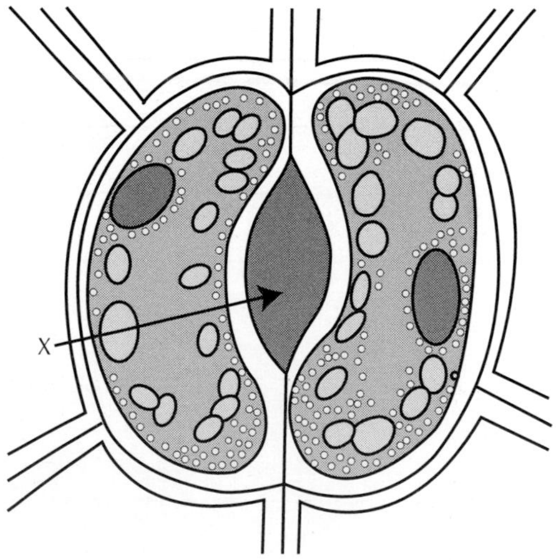

Name structure X, through which water vapour leaves the plant during transpiration.

______________________________ **1**

(d) Suggest a possible benefit of transpiration to a barley plant.

______________________________________ **1**

Total marks 5

MARKS
Do not write in this margin

11. The charts below show the occurrence of five species of plants in samples taken from an area of grassland and from a path passing through it.

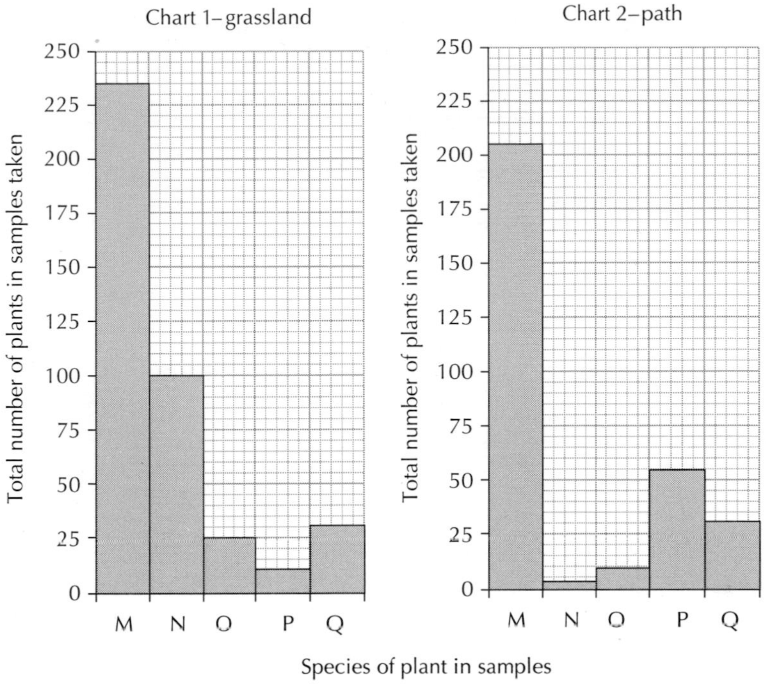

(a) Name **one** method that could be used to sample plants in grassland and describe its use.

Name ________________________ 1

Description of use ________________________

________________________ 1

(b) Give the number of plants of species M that were found in samples taken from the path.

________ plants 1

(c) Describe the effects on the numbers of species O and P of being walked over by people using the path.

Species O ________________________ 1

Species P ________________________ 1

(d) Give the species that is least affected by being walked over.

Species ________ 1

Total marks **6**

12. A group of students set four pitfall traps in a woodland to sample the leaf litter invertebrates living there. The traps were left set for the same length of time.

The table below shows the number and types of invertebrates found.

Invertebrates	**Number**
Woodlice	50
Snails	5
Centipedes	15
Beetles	35

(a) Use the information in the table to complete the pie chart below.

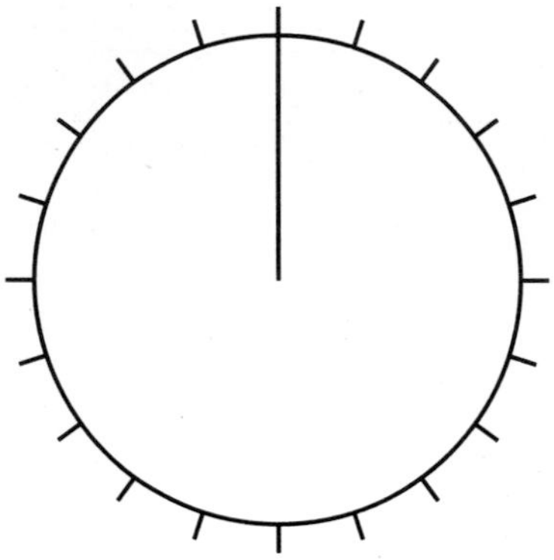

2

(b) Give **one** source of error which can arise when setting up a pitfall trap and suggest **one** method of minimising it.

Source of error ______________________________

______________________________ **1**

Method of minimising error ______________________________

______________________________ **1**

Total marks 4

MARKS
Do not write in this margin

13. The graph below shows the change in plant growth per hectare when different masses of nitrate were added to fields growing an identical variety of crop plant.

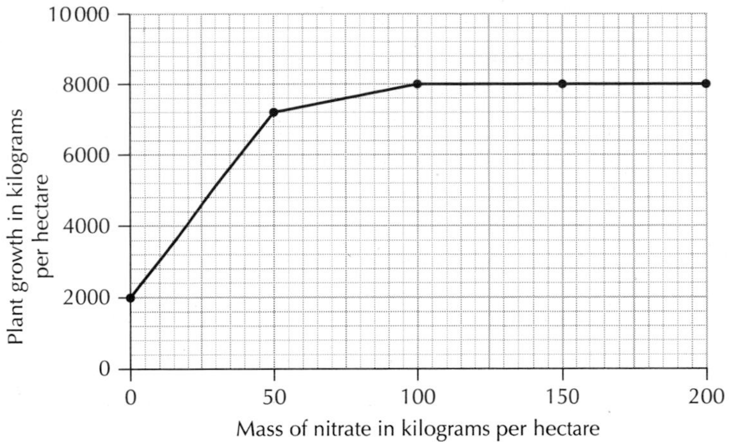

(a) Give the mass at which the concentration of nitrate ceases to be a limiting factor in the growth of the crop plant.

_______________ kilograms per hectare **1**

(b) Calculate the percentage increase in plant growth when the mass of nitrate is increased from 0 to 100 kilograms per hectare.

Space for calculation

_______________% **1**

(c) Name **one** type of substance that is produced by plants using nitrate absorbed from the soil.

_______________________________________ **1**

(d) Name the nitrate-containing substances that are added to soil by farmers to increase the yield of their crops.

_______________________________________ **1**

Total marks 4

MARKS
Do not write in this margin

14. (a) The table below refers to mutation.

Decide if each statement in the table is true or false and tick (✔) the appropriate box.

Statement	***True***	***False***
Mutation is a non-random event.		
Mutation can confer an advantage to an organism.		
Mutation is the only source of new alleles.		

2

(b) Describe how natural selection is involved in the evolution of new species.

____

____

____ **3**

Total marks 5

MARKS
Do not write in this margin

15. Lichens live on the surfaces of walls and trees, and are sensitive to sulfur dioxide, a gas linked with air pollution.

The graph below shows the results of a study in which the percentage of surfaces that were covered by lichens along a line from the centre of a large city that had air polluted by sulfur dioxide was estimated.

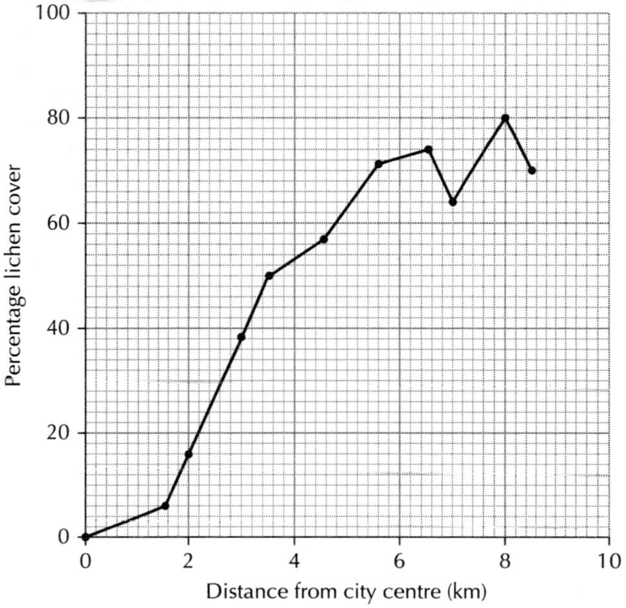

(a) Lichens can be used as indicators of air pollution.

Describe how this statement is supported by the data shown in the graph.

____

____ **1**

(b) At what distance from the city centre was the air pollution the lowest as indicated by percentage lichen cover?

__________ km **1**

(c) As well as sulfur dioxide, polluted air often contains tiny black soot particles.

Predict how these particles would affect the rate of photosynthesis in plants growing in polluted air. Explain your answer.

Prediction ______________________________________ **1**

Explanation _____________________________________

____ **1**

Total marks 4

MARKS
Do not write in this margin

16. Read the following passage and answer the questions based on it.

Blood oxygen levels during exercise

A study was carried out to compare the oxygen saturation of blood taken from muscles in different states of activity. The states of activity included lying down, walking on a treadmill at 1 mile per hour, and jogging on a treadmill at 5 miles per hour.

In order to measure oxygen saturation, catheters with an oxygen sensor were inserted into the left femoral artery and left femoral vein, which transport arterial and venous blood to and from the left leg, respectively. The catheters continuously sampled the blood and measured the oxygen saturation of the blood.

Oxygen saturation is equal to how much oxygen the haemoglobin in the blood is carrying, as a percentage of how much it could theoretically carry. A higher oxygen saturation value implies a higher concentration of oxygen in the blood.

The measurements of the oxygen saturation in the blood taken from the leg muscles in different states of activity are shown in the table below.

Oxygen saturation of blood sample (%)	***Activity***		
	Lying down	***Walking***	***Running***
Femoral artery	97	96	97
Femoral vein	81	68	65

(a) Calculate how many times greater the decrease in oxygen saturation was after running compared to lying down.

Space for calculation

__________________ **1**

(b) The femoral artery transports oxygenated blood to the muscle cells in the legs. Explain why there is a greater decrease in the oxygen saturation in the blood of the femoral vein after running.

____

____ **1**

(c) Give **one** structural difference between an artery and a vein.

____

____ **1**

MARKS
Do not write in this margin

(d) Name the blood vessel through which gas exchange takes place between the blood and the leg muscles.

____ 1

Total marks 4

[END OF QUESTION PAPER]

ADDITIONAL GRAPH PAPER

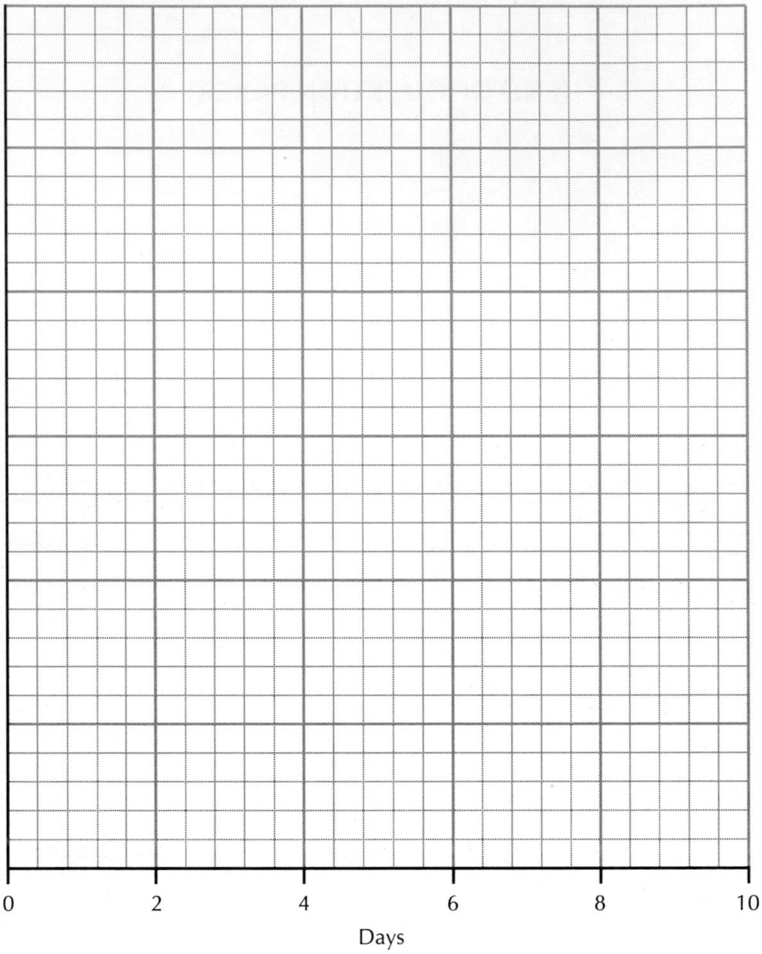

Practice Paper C

SECTION 1 ANSWER GRID

Mark the correct answer as shown ✓

	A	B	C	D
1	○	○	○	○
2	○	○	○	○
3	○	○	○	○
4	○	○	○	○
5	○	○	○	○
6	○	○	○	○
7	○	○	○	○
8	○	○	○	○
9	○	○	○	○
10	○	○	○	○
11	○	○	○	○
12	○	○	○	○
13	○	○	○	○
14	○	○	○	○
15	○	○	○	○
16	○	○	○	○
17	○	○	○	○
18	○	○	○	○
19	○	○	○	○
20	○	○	○	○
21	○	○	○	○
22	○	○	○	○
23	○	○	○	○
24	○	○	○	○
25	○	○	○	○

N5

N5 Biology

Practice Papers for SQA Exams

Practice Paper C
Section 1

Fill in these boxes and read what is printed below.

Full name of centre	Town

Forename(s)	Surname

Try to answer ALL of the questions in the time allowed.

You have 2 hours and 30 minutes to complete this paper.

Write your answers in the spaces provided, including all of your working.

Leckie × Leckie
Scotland's leading educational publishers

SECTION 1 – 25 marks
Attempt ALL questions

1. Red blood cells were placed into a salt solution more concentrated than blood plasma.

Which word best describes the predicted appearance of the cells after a few seconds in this solution?

A Burst

B Plasmolysed

C Turgid

D Shrunken.

2. The diagram below shows molecules present in the cell membrane.

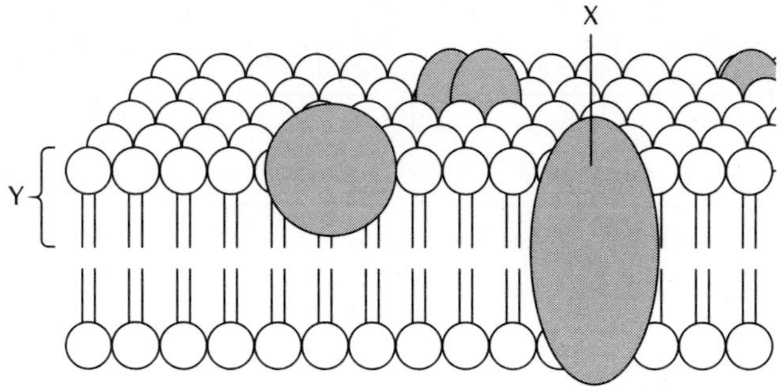

Which line in the table identifies correctly molecules X and Y?

	Molecule X	*Molecule Y*
A	protein	phospholipid
B	protein	cellulose
C	phospholipid	protein
D	phospholipid	cellulose

3. The chart below shows the concentrations of ions in the root cells of a wheat plant and in the soil water in which it is growing.

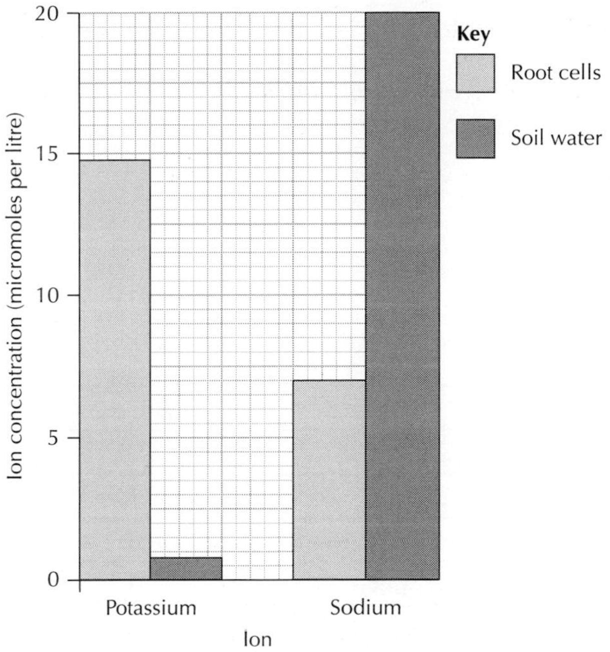

Which one of the following statements is **true**?

A Potassium ions must be taken up from soil by active transport

B Potassium ions can be taken up from soil by diffusion

C Sodium ions can pass out of the root cells by diffusion

D Sodium ions must be taken up from soil by active transport.

4. The diagram below shows a stage in the formation of a molecule of messenger RNA (mRNA).

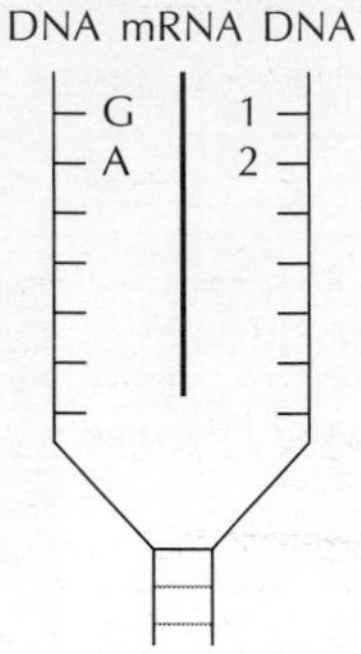

Which line in the table below shows letters that identify correctly Bases 1 and 2?

	Base 1	***Base 2***
A	C	A
B	G	A
C	C	T
D	G	T

5. The graph below shows the effect of pH on the activity of four human digestive enzymes.

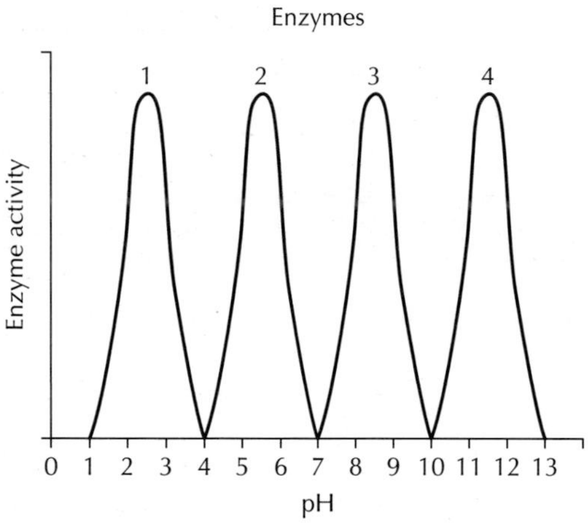

Which enzyme(s) work best in acid pH?

A 1 only

B 1 and 2

C 3 and 4

D 4 only.

6. Which line in the table below correctly identifies the location of the start and the completion of the respiration pathways shown?

	Fermentation pathway		***Aerobic pathway***	
	starts in	***completed in***	***starts in***	***completed in***
A	mitochondria	cytoplasm	mitochondria	cytoplasm
B	mitochondria	mitochondria	mitochondria	cytoplasm
C	cytoplasm	cytoplasm	cytoplasm	mitochondria
D	cytoplasm	mitochondria	cytoplasm	mitochondria

7. Which of the following shows the fermentation pathway in animal cells?

A pyruvate → lactate

B lactate → pyruvate

C pyruvate → ethanol

D lactate → ethanol.

8. Which row in the table below shows properties of stem cells from different locations?

	Location of stem cells			
	In the early embryo		***In the adult body***	
	Can self-renew	***Can specialise***	***Can self-renew***	***Can specialise***
A	✓	✗	✓	✓
B	✓	✓	✓	✓
C	✓	✓	✗	✓
D	✗	✓	✓	✗

9. The list below shows levels of organisation in the body of a mammal.

1 Organ

2 Cell

3 System

4 Tissue

Which is the correct hierarchy into which these levels can be arranged?

A 2→4→3→1

B 4→2→1→3

C 4→2→3→1

D 2→4→1→3

10. The flow chart below shows information about the regulation of blood glucose in humans.

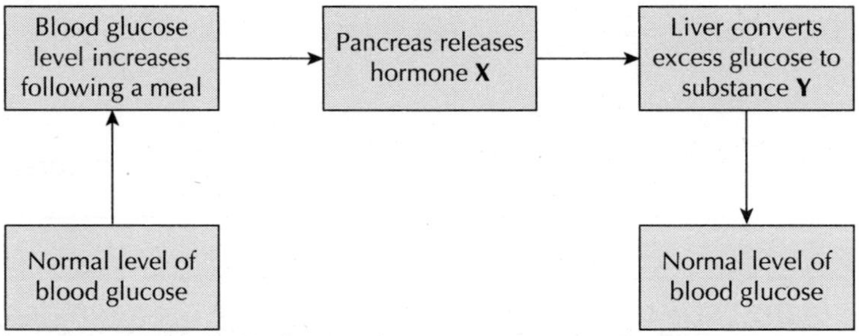

Which line in the table below identifies correctly hormone **X** and substance **Y**?

	Hormone X	***Substance Y***
A	insulin	starch
B	glucagon	glycogen
C	insulin	glycogen
D	glucagon	starch

11. The graph below shows the blood glucose concentration of a patient after he had taken a glucose drink.

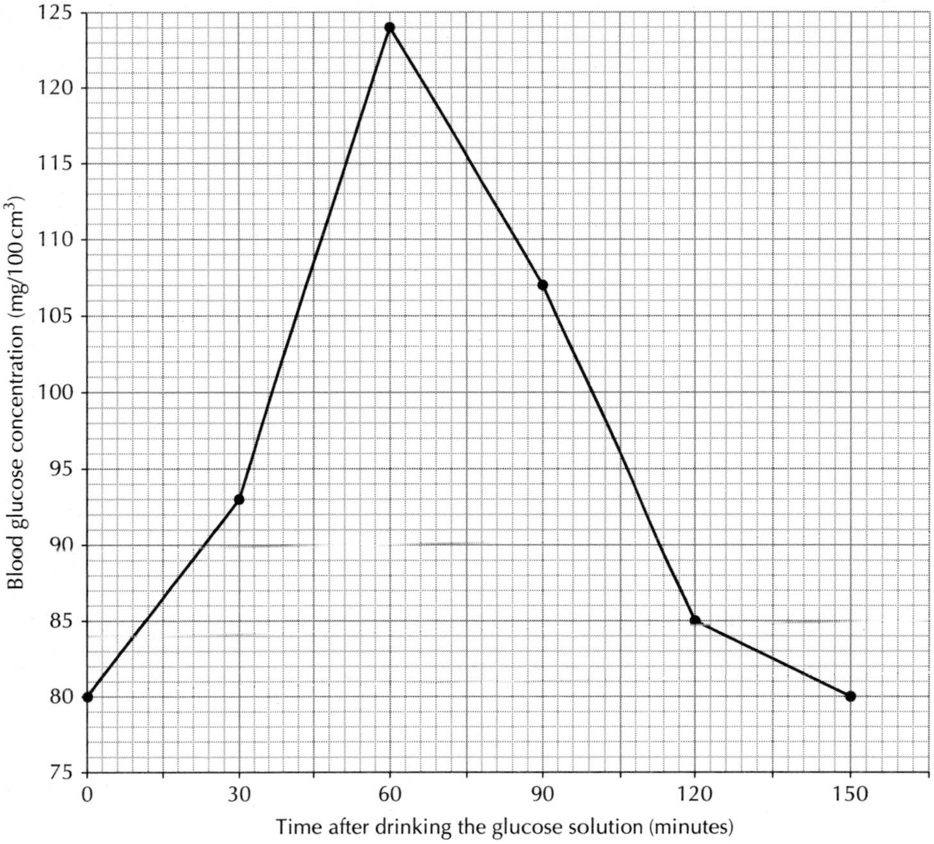

What is the percentage increase in the patient's blood glucose concentration 60 minutes after taking the drink?

A 5.5%

B 44%

C 55%

D 80%

12. The diagram below shows a vertical section through a flower of the pea family.

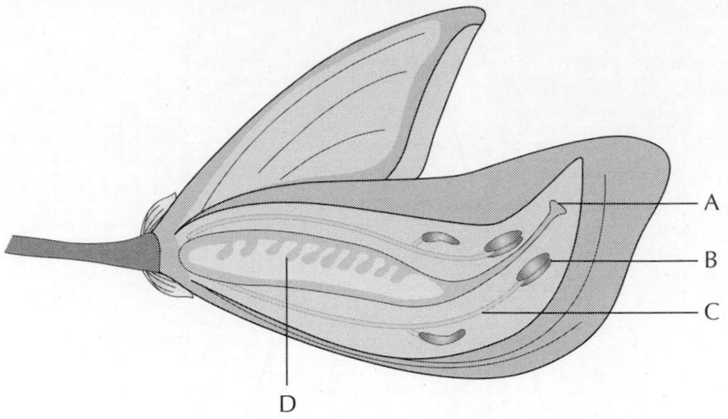

In which structure are female gametes produced?

13. A pea plant with yellow seeds was crossed with a pea plant with green seeds. All of the F_1 plants produced had yellow seeds.

The genotype of the parent plant with green seeds could be described as

A heterozygous and recessive

B homozygous and dominant

C heterozygous and dominant

D homozygous and recessive.

Questions 14 and 15 refer to the diagrams below which show tissues from a plant.

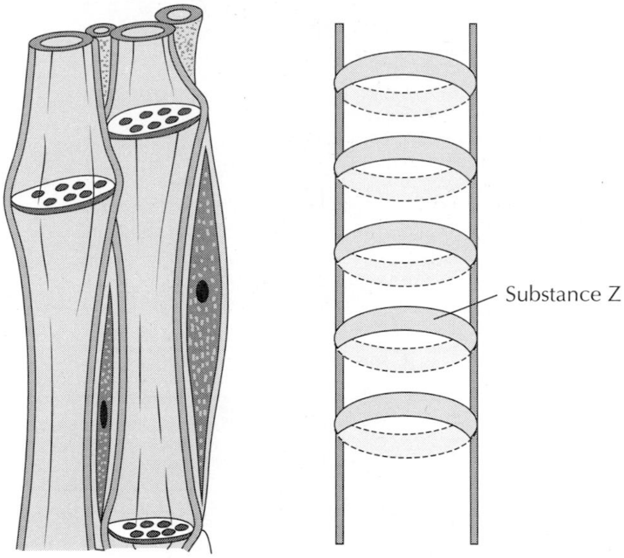

Tissue X　　　　Tissue Y

14. Which line in the table below correctly identifies these tissues?

	Tissue X	*Tissue Y*
A	phloem	xylem
B	phloem	palisade
C	xylem	phloem
D	xylem	palisade

15. What is the function of substance Z?

A To transport water

B To withstand pressure

C To transport sugars

D To trap light.

16. The diagram below shows part of the human digestive system.

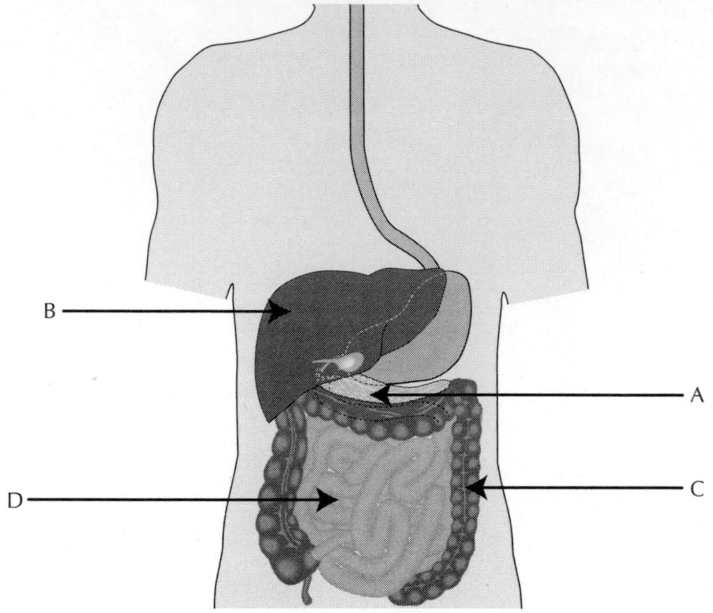

In which region of the diagram would villi be found?

17. Which term describes all the organisms living in an area and the non-living factors with which the organisms interact?

A Habitat

B Ecosystem

C Community

D Niche.

18. The paired-statement key below can be used to identify duckweed plants which grow on the surface of still or slow-moving water.

1	Has roots	go to 2
	No roots	*Wolffia arrhizia*
2	Leaf flat	go to 3
	Leaf-domed	*Lemna gibba*
3	Leaves pointed	*Lemna trisulca*
	Not pointed tip	go to 4
4	Many roots	*Spirodela polyrhiza*
	One root	go to 5
5	Leaves pale grey-green	*Lemna minuta*
	Leaves bright yellow-green	*Lemna minor*

Which duckweed species is shown in the drawing below?

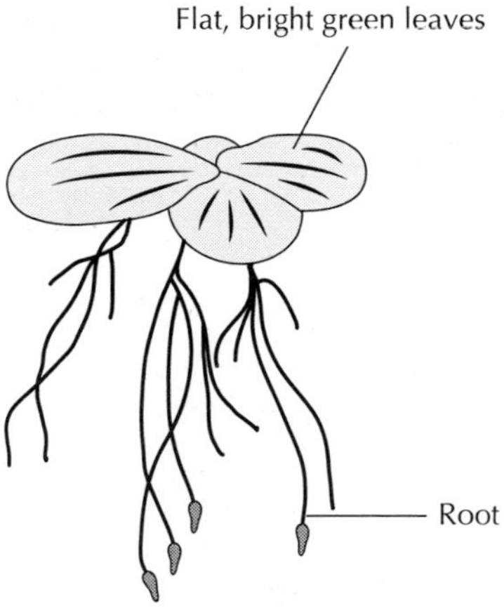

A *Wolffia arrhizia*

B *Lemna trisulca*

C *Spirodela polyrhiza*

D *Lemna minor.*

19. The diagram below shows a pyramid of numbers representing a food chain.

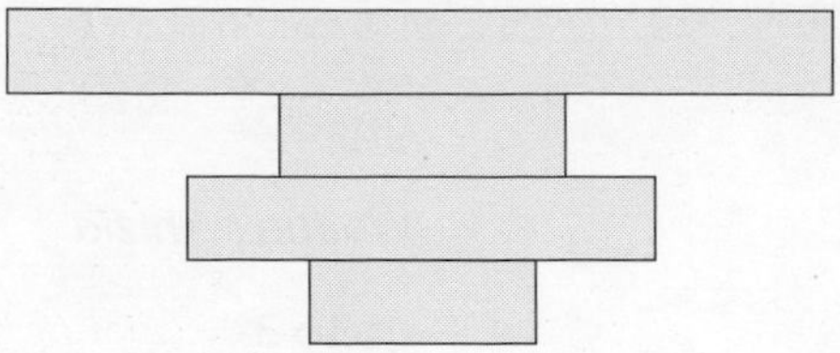

Which of the following food chains could be represented by this pyramid?

A oak tree → moth caterpillar → blue tit → feather mite (a parasite)

B oak tree → greenfly → ladybird → blue tit (a predator)

C heather → moth caterpillar → meadow pipit → merlin (a predator)

D heather → moth caterpillar → meadow pipit → feather mite (a parasite).

20. Some organisms living in seas off the east coast of Scotland are shown in the food web below.

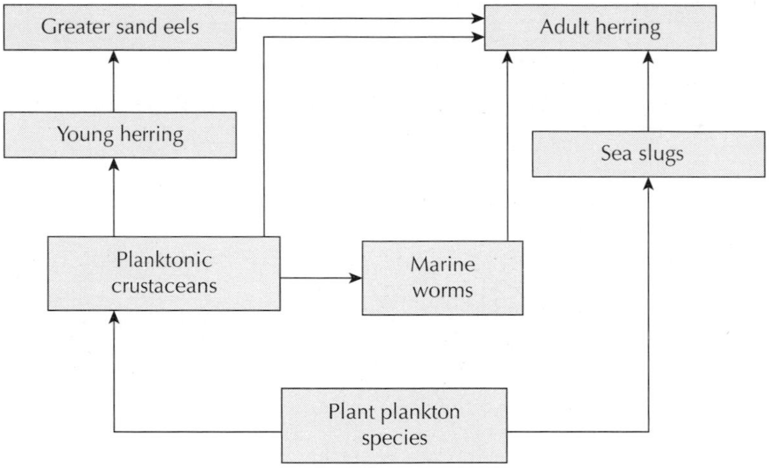

Which line in the table below shows correctly pairs of organisms that are involved in the types of competition shown?

	Type of competition	
	interspecific	***intraspecific***
A	planktonic crustaceans and sea slugs	young and adult herring
B	young and adult herring	greater sand eels and planktonic crustaceans
C	young and adult herring	marine worms and sea slugs
D	planktonic crustaceans and sea slugs	two species of plant plankton

21. The diagram below shows two events in the first stage of photosynthesis in a leaf cell.

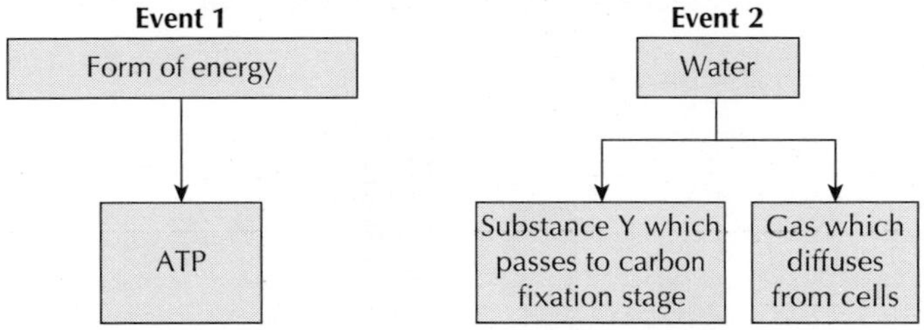

Which line in the table below identifies correctly the form of energy in event 1 and substance Y?

	Form of energy	*Substance Y*
A	light	oxygen
B	chemical	oxygen
C	light	hydrogen
D	chemical	hydrogen

22. The graph below shows the effect of increasing light intensity on the rate of photosynthesis at different temperatures and carbon dioxide concentrations.

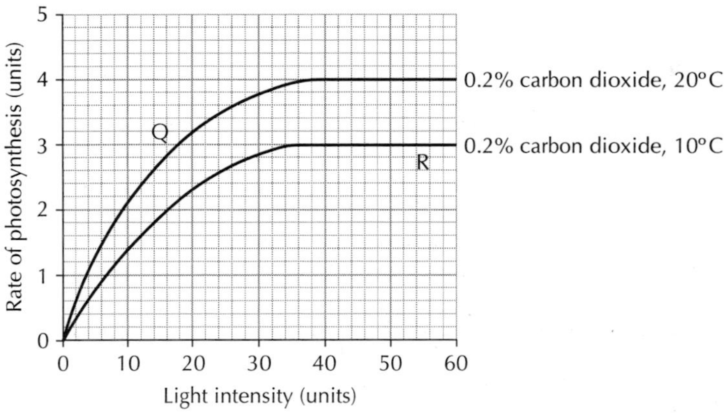

Which line in the table below shows correctly the factors that are limiting photosynthesis at points Q and R on the graph?

	Q	*R*
A	light intensity	temperature
B	carbon dioxide concentration	temperature
C	light intensity	carbon dioxide concentration
D	temperature	light intensity

23. In an investigation, the average numbers of individuals of two forms of the peppered moth in city woodland were estimated every year over a five-year period.

The results are shown on the bar chart below.

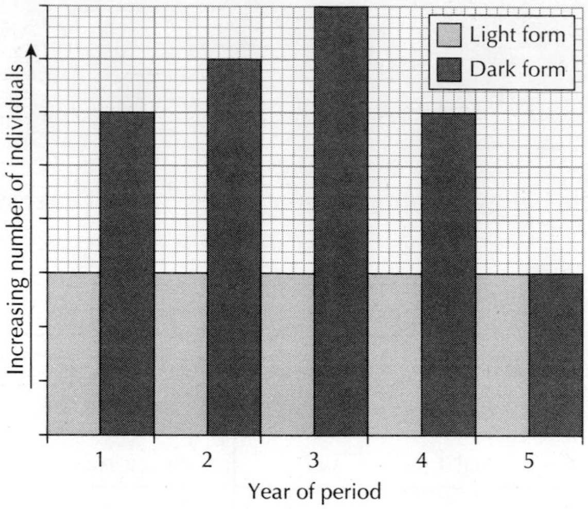

Between which two years of the period did the greatest change in the ratio of light to dark moths occur?

A 1 and 2

B 2 and 3

C 3 and 4

D 4 and 5

24. Plants absorb nitrates from the soil.

Which substances are produced from these nitrates by plant cells?

A Sugars

B Amino acids

C Proteins

D Phosophopholipids.

25. In an investigation, the concentration of a pesticide in the bodies of four individual birds found dead in a farmland area was measured. Two of the birds were predators, and two were prey species.

The results are shown in the table below.

Bird species	***Predator or prey species***	***Concentration of pesticide* (units per gram of muscle)**
wood pigeon	prey	4
skylark	prey	2
sparrowhawk	predator	26
barn owl	predator	16

What is the difference between the average units of pesticide per gram of muscle in the prey species compared to the average in the predator species?

A 12

B 18

C 36

D 39

N5

N5 Biology

Practice Papers for SQA Exams

Practice Paper C
Section 2

Fill in these boxes and read what is printed below.

Full name of centre

Town

Forename(s)

Surname

Try to answer ALL of the questions in the time allowed.

You have 2 hours and 30 minutes to complete this paper.

Write your answers in the spaces provided, including all of your working.

Leckie × Leckie
Scotland's leading educational publishers

SECTION 2 – 75 marks
Attempt ALL questions

1. The diagram below represents a cell from a green plant.

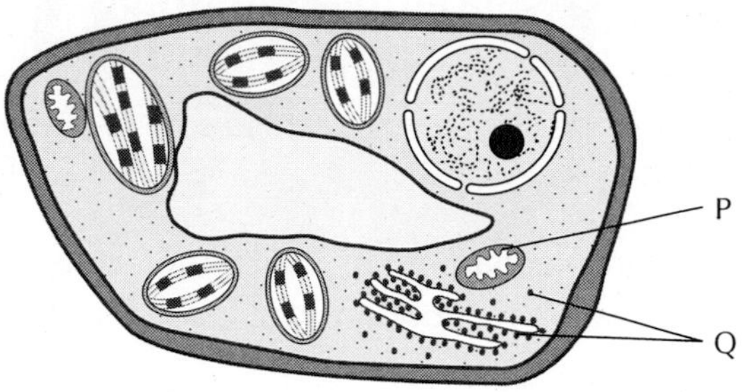

(a) Give evidence from the diagram that suggests that this cell can carry out photosynthesis.

____ **1**

(b) Give the function of structure P.

____ **1**

(c) Name structures Q.

__________________________ **1**

(d) Give **one** structural difference that would be expected between this cell and a fungal cell.

____ **1**

Total marks 4

MARKS Do not write in this margin

2. The diagram below shows the transport of molecule S through a cell membrane.

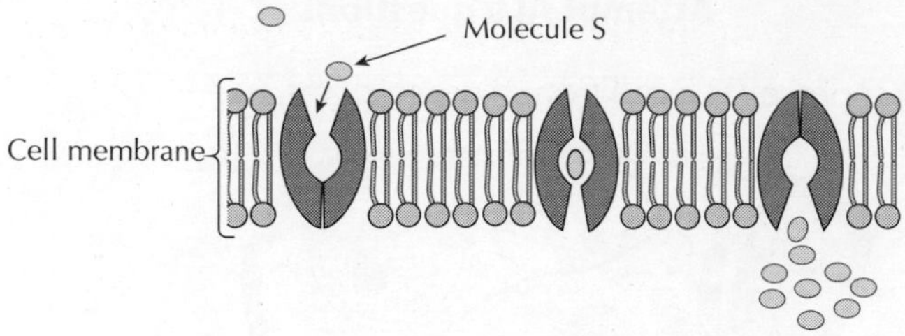

(a) Name the method shown in the diagram by which molecule S is being moved across the membrane. Give **one** reason for your answer.

Method ______________________________ 1

Reason ______________________________ 1

(b) In an investigation, small pieces of tissue of known mass were taken from a water plant submerged in pond water. They were placed into different concentrations of sucrose solution for one hour. After this time, the mass of each piece of tissue was re-measured and expressed as a percentage of its original volume.

The results are shown in the table below.

***Concentration of sucrose solution* (grams per litre)**	***Final mass of tissue* (% of its mass in pond water)**
0	100.0
5	98.5
10	95.0
15	92.5
20	90.5
25	90.0

MARKS
Do not write in this margin

(i) On the grid below, complete the vertical axis and plot a line graph to show the effect of sucrose concentration on the mass of the water plant tissue. (A spare grid, if required, can be found at the end of the practice paper).

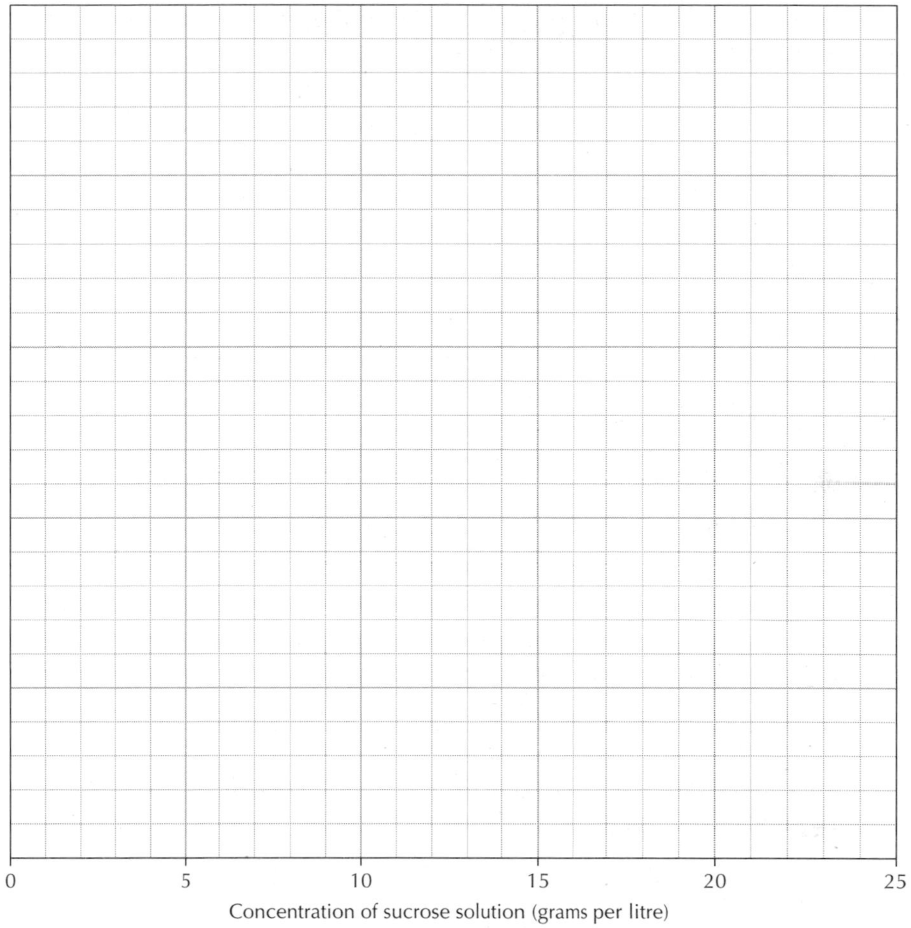

2

(ii) Name the process that causes the mass changes in the water plant tissue.

____ 1

(iii) Using the information available in the table, predict the final mass of a piece of water plant tissue with a starting mass of 2.0 g after it has been immersed in a 25% sucrose solution for one hour.

Space for calculation

Final mass = ____________ g 1

Total marks 6

3. The diagram below shows the genetic modification of a bacterial cell by the transfer of a human gene.

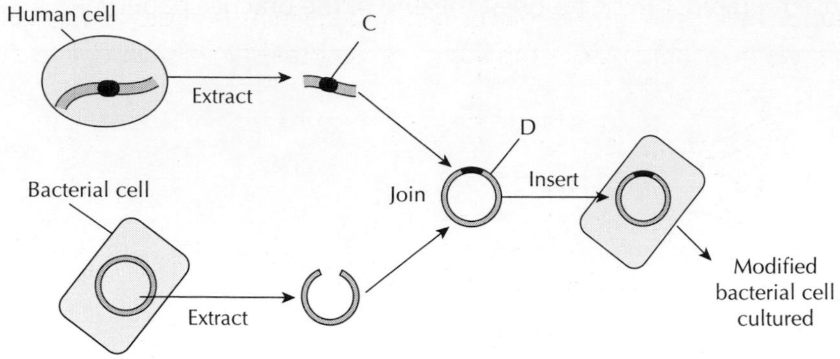

(a) Name the substance of which the human gene C is composed. **1**

____

(b) Identify structure D. **1**

____

(c) Suggest how enzymes are involved in the genetic modification process. **1**

____

____

(d) Give **one** example of a substance produced by the expression of a human gene that has been obtained by this method. **1**

____

Total marks 4

MARKS
Do not write in this margin

4. When mammalian muscle tissue contracts, it decreases in length.

The diagram below shows the procedure involved in an investigation into the effect of different solutions on the lengths of pieces of mammalian muscle tissue. Each piece of muscle tissue was measured before and after five minutes of immersion in the solutions.

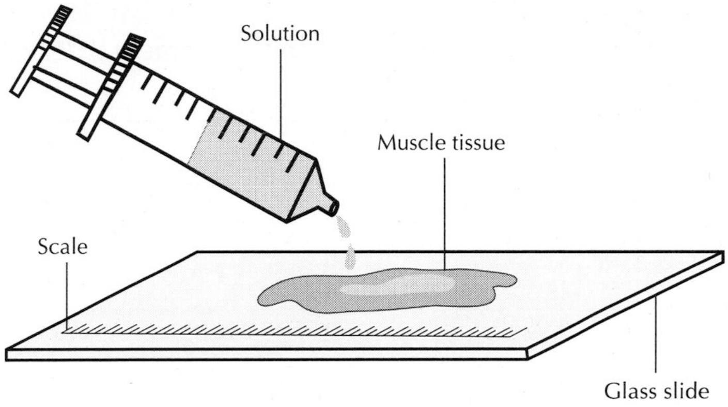

The results are shown in the table below.

Muscle tissue	Solution	Length of muscle tissue (mm)			Percentage difference in length (%)
		at start	after five minutes	difference in length	
A	1% glucose	45	45	0	0
B	1% ATP	50	46	4	
C	distilled water	48	48	0	0

(a) Complete the table by calculating the percentage decrease in length of muscle tissue B.

_______________________% **1**

MARKS Do not write in this margin

(b) Explain why glucose has no effect on muscle tissue A, whereas ATP causes muscle tissue B to contract.

____

____ **2**

(c) Describe why muscle tissue C was included in the experimental design.

____

____ **1**

(d) State what is meant by the term **tissue** in this example.

____ **1**

Total marks 5

MARKS
Do not write in this margin

5. The diagrams below represent stages of mitosis in a plant cell.

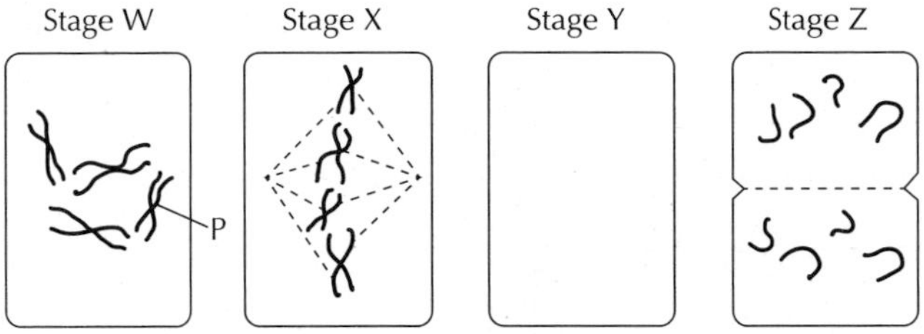

(a) Name the genetic material of which structure P is composed.

____ **1**

(b) Describe events which would occur at Stage Y.

____

____ **2**

(c) Give **one** use of mitosis in organisms.

____ **1**

Total marks 4

MARKS
Do not write in this margin

6. The diagram below shows some structures involved in an example of a reflex action in humans.

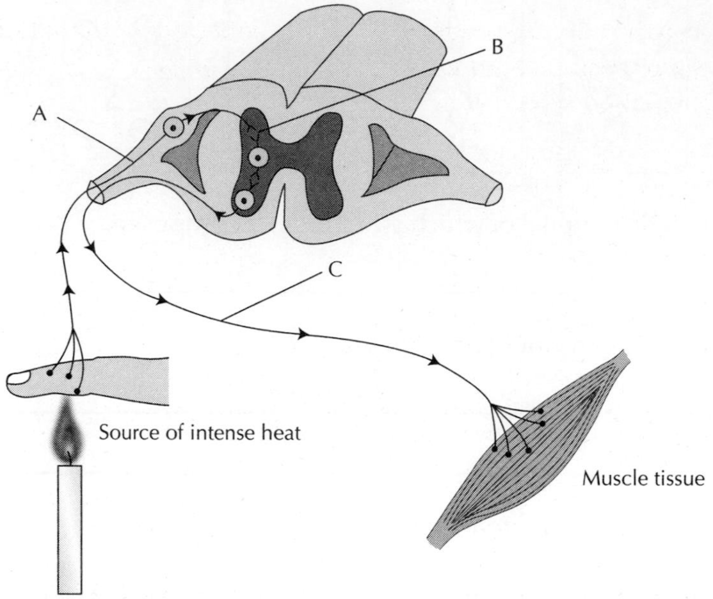

(a) Neurons A, B and C form the reflex arc.

Name each of these neurons.

A ______________________________

B ______________________________

C ______________________________ **2**

(b) Identify the stimulus and describe the expected response in this example.

Stimulus ______________________________ **1**

Description of response ______________________________

______________________________ **1**

(c) Explain the importance of reflex actions in general.

____

______________________________ **1**

Total marks 5

7. Garden pea plants that carry the allele **T** have a tall phenotype.

Plants with the genotype **tt** are dwarf.

20 seeds of a tall variety and 20 seeds of a dwarf variety were germinated and grown for 15 weeks in a greenhouse. After this time the height of each plant was measured, and the results are shown in the charts below.

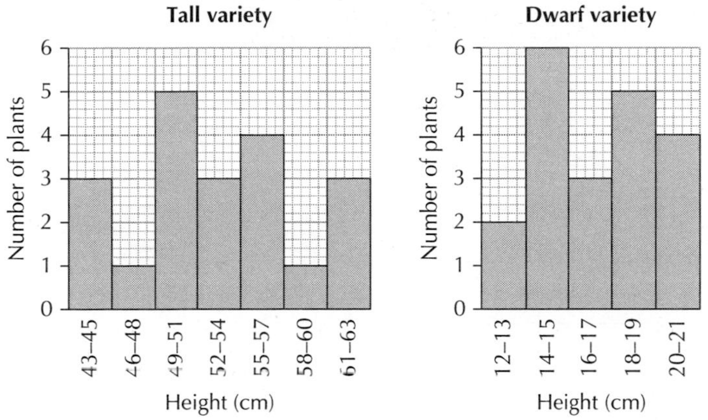

(a) Other than growing the same number of plants for the same time, give **two** variables that should have been kept constant to ensure that comparison of the two varieties was valid.

Variable 1 ____________________ **1**

Variable 2 ____________________ **1**

(b) Give the range of heights in the tall variety.

____________________ **1**

(c) For the information provided, give evidence that shows that height in pea plants shows both discrete **and** continuous variation.

Evidence for discrete variation ____________________

____________________ **1**

Evidence for continuous variation ____________________

____________________ **1**

Total marks 5

8. Dimples are human facial features. Their presence is controlled by alleles of a single gene. The dominant allele (D) gives dimples and the recessive allele (d) gives no dimples.

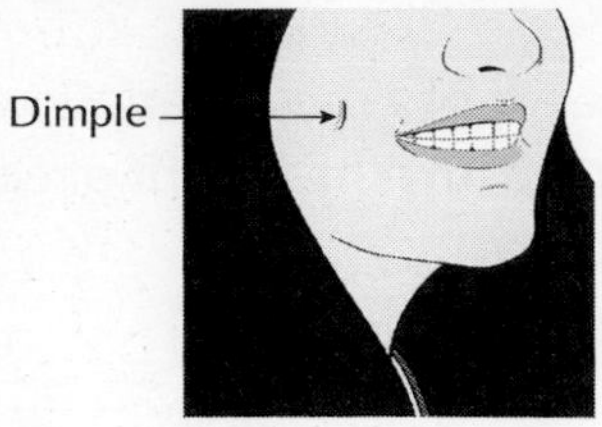

The diagram below shows the inheritance of dimples in a family.

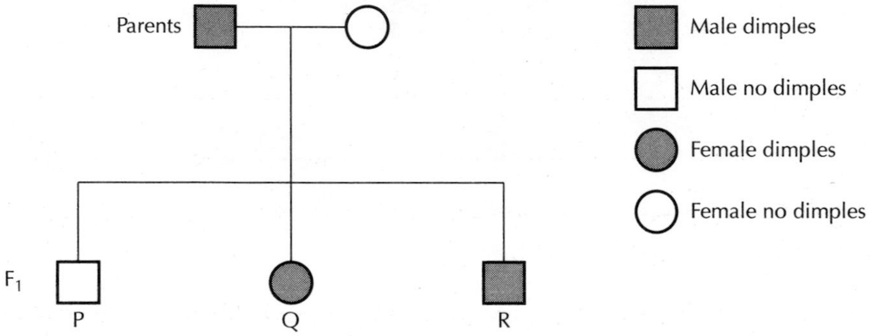

(a) Give the genotypes of the following individuals.

(i) the female parent

____________________ **1**

(ii) son R

____________________ **1**

(b) (i) Daughter Q and a man with no dimples are expecting a baby. Calculate the chance that their baby will inherit dimples.

Space for calculation

____________ % **1**

(ii) Give **one** reason to explain why phenotype ratios among offspring are not always achieved.

____________________ **1**

Total marks 4

MARKS
Do not write in this margin

9. (a) The diagram below shows an external view of the human heart.

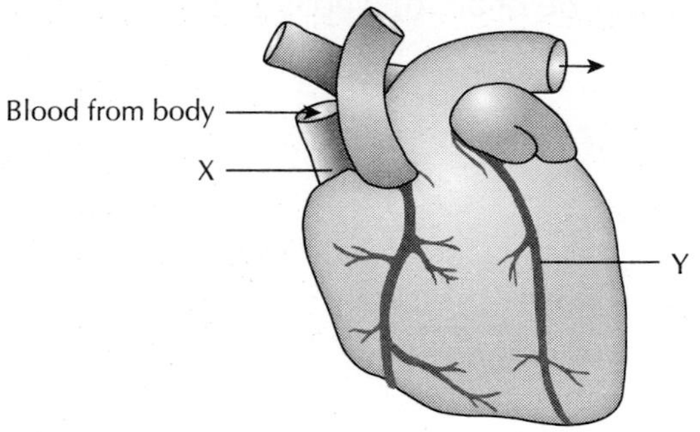

(i) Identify blood vessels X and Y.

X ______________________________ 1

Y ______________________________ 1

(ii) Decide if each of the statements about blood vessels in the grid below is true or false and tick (✔) the correct box.

If the statement is false write the correct word in the box to replace the word underlined in the statement.

Statement	***True***	***False***	***Correction***
Arteries carry blood from the heart.			
Veins exchange materials with the tissues.			
Capillaries have valves.			

2

(b) The graph below shows the effect of carbon dioxide concentration in the air on the volume of air inhaled into the lungs of an individual at rest.

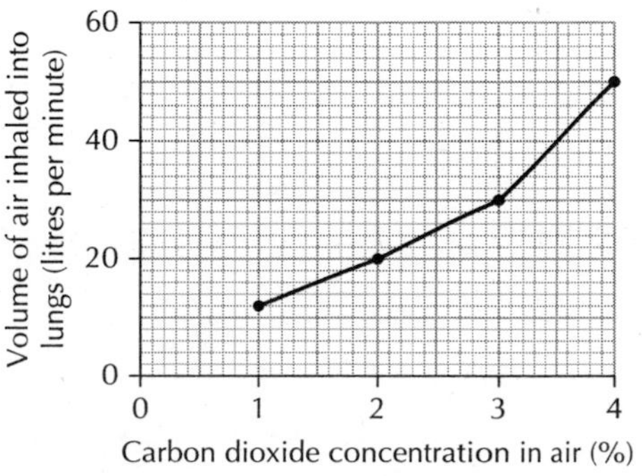

MARKS
Do not write in this margin

(i) Calculate the volume of carbon dioxide entering the individual's lungs each minute when the volume of air inhaled is 20 litres per minute.

Space for calculation

____________________ litres **1**

(ii) Calculate the increase in volume of air entering the lungs per minute when the concentration of carbon dioxide in the air increases from 1% to 4%.

Space for calculation

____________________ litres **1**

Total marks 6

MARKS
Do not write in this margin

10. In an investigation into the effects of temperature on the rate of transpiration in a leafy seedling, the potometer below was set up.

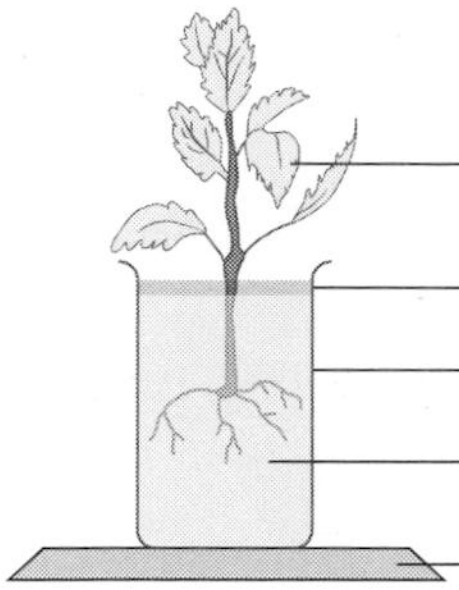

Transpiration rate was measured at different temperatures. The results are shown in the table below.

***Temperature* (°C)**	***Transpiration rate* (grams of water per cm^2 of leaf per minute)**
10	0.2
15	0.3
20	0.4
25	0.5

(a) Identify the observations or measurements that would have to be made to obtain the values for the rate of transpiration shown in the table.

____

____

____ **3**

(b) The following factors can affect transpiration rate in plants.

Light intensity **Atmospheric humidity** **Air movements**

Choose **one** of these factors and describe how the apparatus above could be modified to investigate this factor.

Factor ______________________

Description ______________________________________

____ **2**

Total marks 5

MARKS
Do not write in this margin

11. The pie chart below shows one estimate of the percentage of the Earth's land area occupied by different ecosystems.

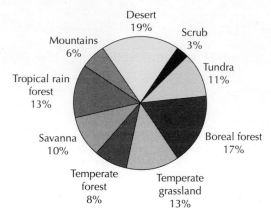

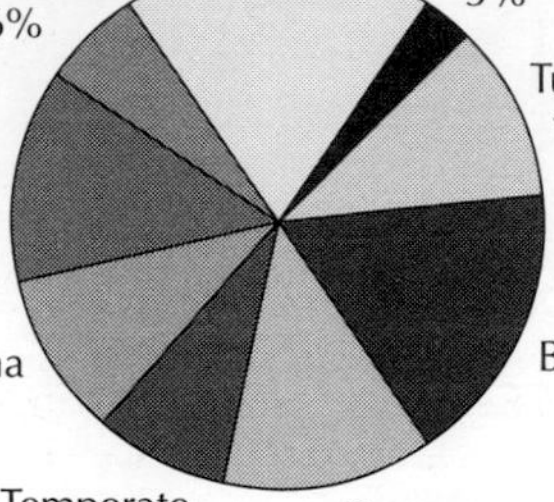

(a) Desert ecosystems are found in many parts of the world.

(i) Apart from the organisms which live there, state what else makes up a desert ecosystem.

________________________________ **1**

(ii) Desert organisms have adaptations for their way of life.

Describe what is meant by an adaptation.

____

________________________________ **1**

(b) Calculate the total percentage of land surface occupied by the various types of forest ecosystems.

Space for calculation

________________% **1**

(c) Within ecosystems each species occupies its own niche.

Describe what is meant by the term niche.

____

________________________________ **1**

(d) The diagram shows a food chain from a temperate forest ecosystem.

The arrows represent energy flow.

oak tree leaves → greenfly → ladybird → blue tit → sparrowhawk

(i) Explain how the oak tree leaves obtain the energy trapped within them.

____

____ **1**

(ii) Give **one** fate of energy present in the greenfly population which does not pass to the ladybird population.

____ **1**

Total marks 6

12. In an investigation to compare the populations of a species of ground beetle living on the soil surface in two different areas of grassland, sampling was carried out using pitfall traps.

(a) Give **two** precautions that would have to be taken to ensure that the sampling method allowed valid **comparison** of the two areas.

1 ____

2 ____ **2**

(b) Describe a source of error that can arise when using pitfall traps.

____ **1**

(c) During the investigation, a number of abiotic factors related to the soil were also measured.

Name **one** abiotic factor that is related to soil and describe how it could be measured.

Abiotic factor ____________________________ **1**

Method of measurement ____________________________

____ **1**

Total marks 5

13. (a) The apparatus shown below was set up to measure the rate of photosynthesis in pondweed.

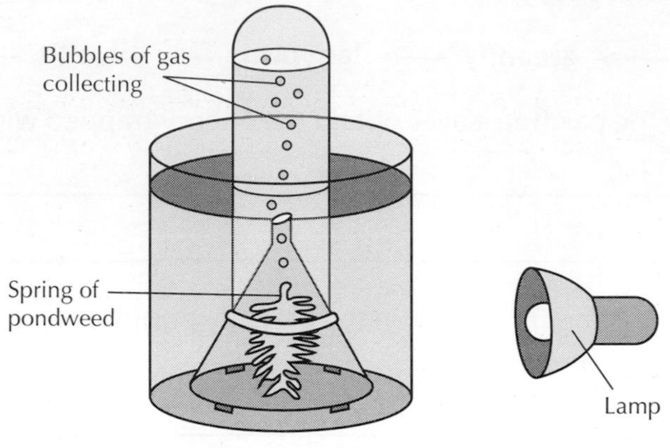

(i) Name the gas present in the collecting bubbles.

______________________________ 1

(ii) Describe how the apparatus could be used to show the effects of light intensity on the rate of photosynthesis.

____

____

______________________________ 3

(b) The graph below shows some results of an experiment to show the effect of carbon dioxide concentration on the rate of photosynthesis at different temperatures.

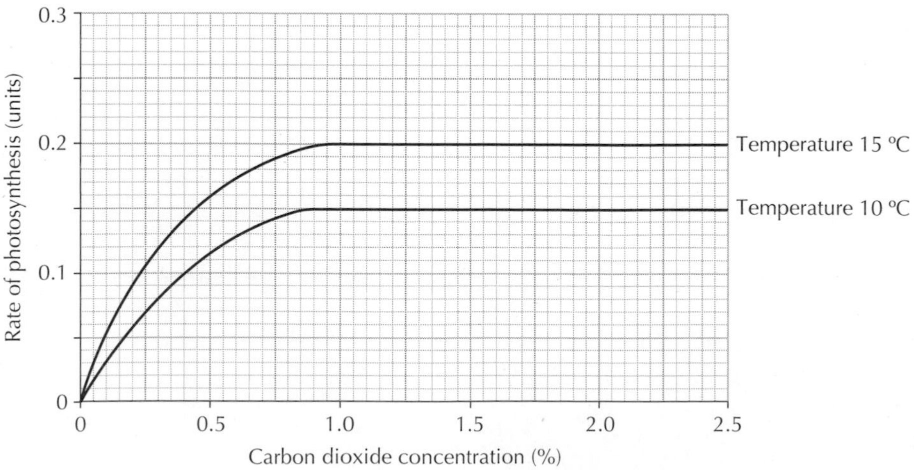

(i) Give the carbon dioxide concentration at which the rate of photosynthesis first reached its maximum at 15 °C.

_______________% **1**

(ii) Calculate the increase in the rate of photosynthesis when the temperature was raised from 10 °C to 15 °C at a carbon dioxide concentration of 2.0%.

Space for calculation

_______________units **1**

(c) Sugar produced by photosynthesis can be converted into starch.
Give the function of starch in green plants.

_______________________________________ **1**

Total marks 7

MARKS Do not write in this margin

14. Read the following passage and answer the questions based on it.

Genetically modified (GM) rice

Nitrogen is the most important soil nutrient for plants and a major factor which can limit crop productivity. Nitrogen-rich fertilisers are often used to boost crop growth but plants are inefficient at taking up the nitrate from applied fertiliser. As a result, excess nitrates frequently leach from the soil into waterways and cause algal blooms. Dead algae become food for bacteria which consume oxygen needed for healthy aquatic ecosystems.

To meet growing food demands, the global use of nitrate fertiliser increased from 3.5 million metric tonnes in 1960 to 87 million metric tonnes in 2000, and is projected to increase to 249 million metric tonnes by the year 2050.

Scientists in Canada have successfully developed a genetically modified (GM) rice variety which produces enzymes that allow more efficient nitrate uptake than unmodified varieties. This may reduce the need for nitrogen-rich fertilisers and at the same time increase yields.

These crops not only have the potential to lower production costs and reduce environmental pollution, but their increased productively could make a significant contribution to our long-term food security.

(a) Explain why fertiliser use creates excess nitrates in soils.

____ 1

(b) Explain how algal blooms can lead to the de-oxygenation of freshwater ecosystems.

____ 1

(c) Calculate the projected global increase in nitrate fertiliser use between 1960 and 2050.

Space for calculation

______________ metric tonnes 1

(d) Explain why the GM rice described in the passage could reduce the amount of nitrate fertiliser which needs to be applied.

____ 1

(e) Apart from reducing fertiliser application, give **one** other benefits of growing the GM rice.

____ 1

Total marks 5

MARKS
Do not write in this margin

15. On the Galapagos Islands of the Pacific Ocean, speciation has produced a group of similar finch species, as shown in the diagram below. The group arose from a single ancestor species, which reached the islands from the South American mainland millions of years ago.

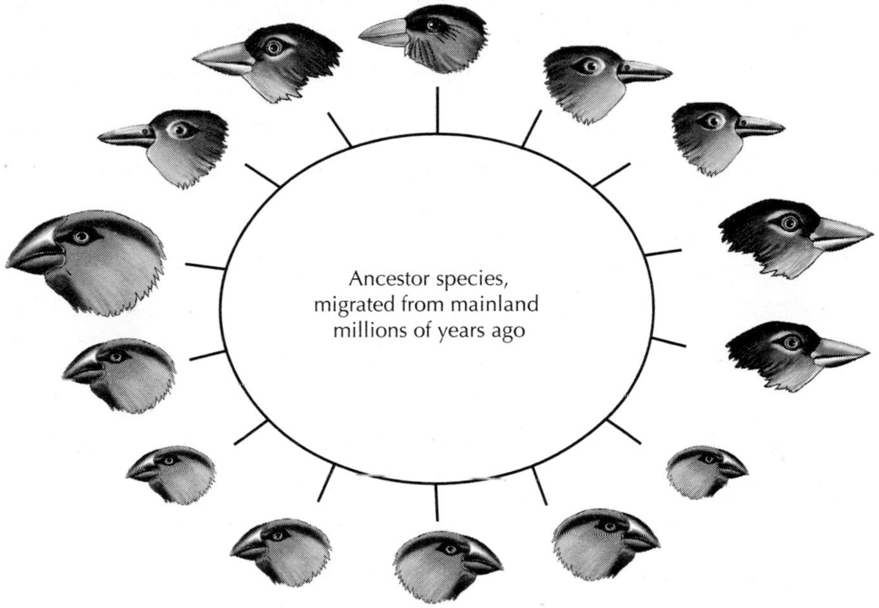

(a) The list below shows processes involved in speciation.

mutation **isolation** **natural selection**

Describe how these processes have led to the production of the group of finch species in the diagram above.

____

____

____

____

____ **3**

(b) Give the term applied to mutations that confer neither advantage nor disadvantage.

____ **1**

Total marks 4

[END OF QUESTION PAPER]

ADDITIONAL GRAPH PAPER

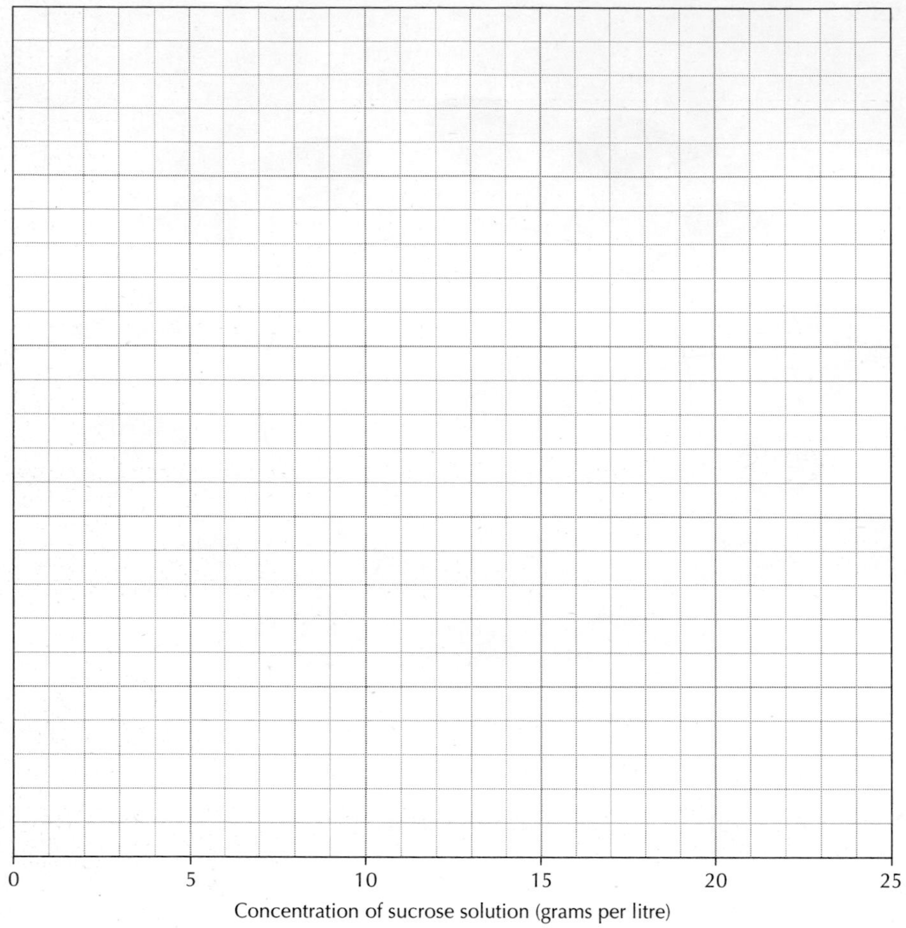

Answers

Answers to Practice Papers

Practice Paper A

Section 1

Question	Response	Mark	Top Tips
1.	B	1	You must be able to identify cell structures from diagrams and know their functions.
2.	C	1	Find the number of cells and then divide it by the total number of cells – then multiply the answer by 100: (15/50) x 100 = 30%
3.	B	1	Remember that cell walls in fungi have a different composition and structure.
4.	D	1	First divide 0.5mm by 500 (the magnification) = 0.001 (the length of this cell) Next divide 1.0mm by 0.001 = 1000
5.	B	1	You need to go through each option – it will take a bit of time and making some sketches might help. You can draw on the question paper!
6.	A	1	You need to know the three features of active transport – against concentration gradient, involves proteins, needs energy/ATP.
7.	D	1	Language is crucial – you need to know the terms **chromatid** and **spindle fibre**, and their roles in mitosis.
8.	B	1	DNA looks like a twisted ladder – the rungs are the bases that carry the genetic code.
9.	D	1	Bacterial chromosomes are usually long and coiled up, but plasmids look neat and circular.
10.	D	1	You need to learn the levels of organisation in biology – cell, tissue, organ, system, organism.
11.	C	1	Watch out for the **bold** in the question – then tick off the true options as you work through them.
12.	A	1	This is a standard type of brain diagram – you must be able to identify where particular functions occur.
13.	C	1	The key here is reading the question – the word **respond** is crucial – candidates could be wrongly attracted by pancreas, which **produces** insulin.

Question	Response	Mark	Top Tips
14.	B	1	You need to appreciate that gametes are the only haploid cells mentioned in National 5 Biology.
15.	C	1	A = B × C so B must be A divided by C. Calculator almost essential! Remember that 4 litres = 4000 cm^3
16.	A	1	Remember transpiration is higher in warmer temperatures but less in humid conditions.
17.	B	1	Find the % loss in mass = 8% then find 8% of 600 and substract that from 600
18.	B	1	You need to learn these words – Why not make yourself a set of flash cards? Put the word on one side and the meaning on the other.
19.	D	1	The **bold** is vital, and it is the term **community** that needs to be learned.
20.	D	1	The word **bio**tic sounds like **bio**logy for a good reason – you are looking for **living** factors.
21.	C	1	A bit tricky but if 90% is lost, 10% is kept. Writing the figures under the organisms' names in the chain will help keep you right.
22.	A	1	Take each option in turn. B and C are clearly wrong but D looks attractive. Remember that plants need nitrates and make their own amino acids and proteins.
23.	A	1	It is vital to remember that the 0 value counts – so the total number of limpets is divided by the 9 quadrats.
24.	A	1	It is worth trying to sort out the difference between validity and reliability – it's tough because they do overlap a bit.
25.	C	1	Mutation is the **only** source of new variation.

Practice Paper A

Section 2

Question			Expected response	Mark	Top Tips
1.	(a)		X: phospholipid = **1** Y: protein = **1**	2	You need to learn the appearance of the two molecules in diagrams.
	(b)		Selectively Down Does not require **All 3 for 2 marks, 2 or 1 for 1 mark**	2	Remember that the concentration gradient is a bit like a physical slope – so, down and up – with and against are good terms!
	(c)	(i)	Place in a solution of lower water concentration than cell sap	1	Position of the cell membrane is vital in identification of cell condition.
	(c)	(ii)	Plasmolysed	1	Remember **PS** – **P**lasmolysed in **S**trong solution.
2.	(a)		P R Q	1	The words **building up** in the stem of the question are crucial to answering. **S**ynthesis **s**tarts with **s**mall molecules.
	(b)		Active site	1	The shape of the active site allows reaction with specific substrate molecules.
	(c)		enzyme substrate complex	1	You are expected to know this term applied to stage in the reaction when the substrate attaches to the active site of the enzyme.
	(d)		active site shape altered/ denatured = **1** cannot bind to substrate = **1**	2	Temperature is critical in biology – proteins don't like heat!
	(e)		pH/enzyme concentration/ substrate concentration	1	The rate of reaction is affected by temperature and pH as mentioned in the Course Specification, but these other answers would also be acceptable.

Question			Expected response	Mark	Top Tips
3.	(a)		Scales and labels = **1** Points and connection = **1**	2	Include zeros and highest values on even scales. Include units with labels. Plot with a sharp pencil. Connect plots with straight lines.
	(b)		Carbon dioxide	1	Remember CO_2 is produced in fermentation as well as in aerobic respiration.
	(c)		Measure volume of gas rather than counting bubbles	1	**Accurate** is the key word – this usually relates to the measurement method.
	(d)		Rate would decrease = **1** Enzymes work slowly in cool conditions = **1**	2	You must realise that fermentation is enzyme-controlled.
4.	(a)	(i)	0.5%	1	Find 75% of the starting value for glucose = 60mM per litre then use a ruler to read down from this value to the ethanol conc. and read across to the right scale = 0.5%
	(a)	(ii)	50 hours	1	Simply read off at the time glucose conc. becomes 0
	(a)	(iii)	0.02% ethanol per hour	1	Find the increase in % ethanol = 4% then divide it by 200 hours.
	(b)		Glucose disappears very quickly from solution = **1** Ethanol continues to increase in concentration after that = **1**	2	You have to notice that ethanol production keeps on increasing even although there is no glucose left in the solution.
	(c)		4.5%	1	Simply read off the value at the end of the fermentation.
5.	(a)		Sensory neuron	1	There are three types of neuron to know – sensory, relay and motor.
	(b)		Synapse = **1** Transfer electrical impulses between neurons = **1**	2	Electrical impulses can only cross when the synapses are filled with chemical transmitter.
	(c)		Protection of skin from excess heat/burning = **1** Protects the body from harm = **1**	2	The word *protection* is vital here, and the diagram in the question gives the clue to the type of damage avoided.

Question			Expected response	Mark	Top Tips
6.	(a)		J: Rr K: Rr L: rr **All 3 for 2 marks, 2 or 1 for 1 mark**	2	It is worth adding the known alleles onto the diagram on the paper to make answering easier.
	(b)		Homozygous	1	Remember the same alleles is homozygous but **different** alleles is **hetero**zygous.
	(c)		N has allele R because he is a roller = **1** The other allele could be either R or r = **1**	2	Doubt about offspring exists because at least one parent is heterozygous.
7.	(a)		W xylem, water/minerals X phloem, sugar **All 4 for 3 marks, 3 or 2 for 2 marks or 1 for 1 mark**	3	Just learn it but you could try the F sounds – *phloem* for *food.*
	(b)		Lignin	1	**XL** – **X**ylem has **L**ignin.
8.	(a)	(i)	Carbon dioxide	1	The only gas waste of respiration.
	(a)	(ii)	Alveolus	1	Don't get confused with villus.
	(a)	(iii)	1: moist surface	1	The moist lining layer is vital but missing from the diagram.
			2: large surface area **OR** thin walls	1	Both of these features are shown although the large surface is not so obvious.
	(b)	(i)	Age Number of cigarettes smoked daily **both.**	1	Look carefully for the factors that vary in the data.
	(b)	(ii)	200%	1	Use a clear plastic ruler to help with the graph reading and remember that doubling a number is a 100% increase.
9.	(a)		Shape – biconcave = **1** Explanation – large surface area = **1** For absorption of oxygen = **1**	3	Shape is visible from the diagram but you need to remember the term used for it.
	(b)		Lymphocyte = **1** Phagocytosis = **1**	2	The phagocyte's name is a clue to its function!

Question			Expected response	Mark	Top Tips
10.	(a)		Quadrats Drop randomly **Both needed for mark**	1	The only plant-sampling technique in National 5 assessment is quadrats.
	(b)		3:1	1	Ensure that your ratio has only whole numbers.
	(c)		20 000	1	This is where the need to know the number of m^2 in a hectare given in the question comes in.
	(d)		Light intensity = **1** Shade of trees in woodland reduces photosynthesis at ground level = **1**	2	You need to visualise a wood compared with an open grassy area – light and shade should come to mind! There could be other answers too, though.
11.	(a)		Leached into the river **OR** washed/drained into the river in rain	1	Notice the word "leach"-the best term to use here.
	(b)		Algae (and other plants) die = 1 Their bodies are decomposed/food for by bacteria = 1 Bacterial numbers increase (greatly) = 1 Bacteria are aerobic/use large quantities of oxygen = 1 **[Any 3]**	3	This is a complex story so you need to spend time learning it.
	(c)		Downstream from/after station 2 the oxygen levels rise/bacterial numbers increase	1	Quite straightforward if you look at the the information in the table.
12.	This type of question is designed to test both skills and knowledge. Read these short passage questions carefully. It is sometimes quite useful to have a quick skim through the questions being asked before reading the passage. For most answers you can give the same phrases used in the actual passage, unless asked for a particular meaning or explanation.				
	(a)	(i)	Respond quickly to habitat variation **OR** easy to identify/count **OR** abundant **OR** active during daytime **Any 2, 1 mark each**	2	All mentioned clearly in the passage.
	(a)	(ii)	Pollution	1	You have met indicators before in the context of pollution.
	(b)		Linnet **OR** goldfinch	1	Taken straight from the passage.

Question			Expected response	Mark	Top Tips
	(c)	(i)	–2.9	1	Get the figures from the passage and do the calculation.
	(c)	(ii)	+18.2	1	
13.	(a)		A = hydrogen B = ATP C = oxygen **All 3 for 2 marks, 2 or 1 for 1 mark**	2	This is a useful diagram–you could copy it and put in the missing information for your revision notes.
	(b)	(i)	Negative/no starch present = **1** CO_2 needed for photosynthesis = **1**	2	Starch storage is a sign that photosynthesis has happened, and excess sugar/glucose has been produced.
	(b)	(ii)	Set up as leaf A but without substance to absorb CO_2	1	Controls allow comparison with results and show if an experimental variable is causing a result.
	(b)	(iii)	Repeat experiment but remove glass tubes, cover one leaf to exclude light	1	This is a common question type – there are three parts to the standard answer. Repeat, hold original variable constant, alter new variable.
14.	(a)		Species Q Correctly adapted beak and habitat preference **both needed for mark**	1	Spotting the link here is crucial – the table describes beak shape, and the diagrams show beak shape.
	(b)		Different food = **1** Different habitat = **1**	2	Interspecific competition occurs when the **same** resources in the **same** habitat are required by **two** species – competition is reduced when requirements are different.
	(c)		R P Q	1	What about trying **I'M** a **N**ew **S**pecies. **I**solation – **M**utation – **N**atural **S**election?

Practice Paper B

Section 1

Question	Response	Mark	Top Tips
1.	B	1	Make sure that you can identify cell structures from diagrams and know their functions.
2.	C	1	Count the cell lengths up the field of view and divide the number into 1000 micrometres.
3.	D	1	The cell membrane has a double layer of phospholipid molecules and a patchy arrangement of proteins.
4.	C	1	Remember that active transport is the movement of molecules against the concentration gradient and requires ATP and protein carriers.
5.	C	1	Candidates sometimes confuse specific and optimum – make sure you know the difference.
6.	B	1	Substrates (P + Q) and enzyme must be present at the start. The enzyme remains unchanged and must be R. Product is T and so S is the enzyme-substrate complex.
7.	B	1	The extraction order does not matter but the insertion order does!
8.	A	1	Important to see that X (pyruvate) is a junction for both pathways.
9.	B	1	Decrease in respiration rate can result from a reduction in respiratory substrate as in option B.
10.	A	1	Stem cells are unspecialised cells that divide by mitosis to self-renew.
11.	C	1	Evidence of dominance often comes from looking at phenotypes in the F_1 – the word *all* in option C is a big clue.
12.	C	1	Rough working will be needed here.
13.	B	1	Be aware that flowers take many different forms but the structures are usually in the same positions relative to each other.

Question	Response	Mark	Top Tips
14.	A	1	Make sure that you can label this type of section through a leaf and give the function of the cells.
15.	C	1	**PS** – **P**hloem carries **S**ugar.
16.	A	1	The middle single un-branched vessel is the lacteal, which leads into the lymph vessels.
17.	D	1	Biotic and biology are related – you are looking for living factors under biotic and non-living ones under abiotic.
18.	C	1	Make sure you understand how the pyramid shape relates to the numbers and energy available at the different levels.
19.	C	1	The stoats have only one other competitor for the rabbits-the foxes, and one other competitor for the voles-the kestrels.
20.	A	1	It might be useful to mark trends onto the table in the question paper.
21.	D	1	It's crucial here to notice that increasing distance decreases light intensity.
22.	A	1	Think about energy forms. Light energy trapped by chlorophyll and chemical energy trapped into ATP.
23.	A	1	Remember – **I'M** a **N**ew **S**pecies – **I**solation **M**utation **N**atural **S**election.
24.	B	1	A two-step calculation – first find the increase, then divide it by the original value and multiply the answer by 100.
25.	D	1	There are two variables affecting the result because the disease requires overcrowded conditions to flourish.

Practice Paper B

Section 2

Question			Expected response	Mark	Top Tips
1.	(a)		Show smaller vacuole = **1** Show cytoplasm/cell membrane pulled from wall = **1**	2	Draw carefully using the shading in cell A as a key for your drawing – sharp pencil needed.
	(b)		Turgid	1	Identifying the membrane is vital in seeing what's going on.
	(c)	(i)	Osmosis	1	What's moving? If it's water, it's by osmosis!
	(c)	(ii)	Does not require (additional) energy	1	Passive is the opposite of active – no additional energy required.
	(d)		**T**issue **S**imilar	1	Cell, tissue, organ, system, organism – levels of organisation again.
2.	(a)		Chromatids pulled apart/move toward poles	1	Terms are vital – you will need **chromatid** and **poles**…
	(b)		Spindle fibre	1	… and now you need **spindle fibres**.
	(c)		Identical Diploid Unspecialised **All 3 for 2 marks, 2 or 1 for 1 mark**	2	These cells have divided by mitosis so they must be genetically identical and diploid. They can go on to become any of the mature cells of the plant so, at this stage, they must be unspecialised.
3.	(a)		DNA	1	DNA carries the genetic code.
	(b)		1: T/thymine = **1** 2: C/cytosine = **1**	2	Bases are in complementary pairs – you just need to learn them!
	(c)		Ribosomes	1	**R**NA goes to **R**ibosomes.
	(d)		The order/sequence of bases	1	The bases are like an alphabet, so their order gives the code its sense.
	(e)		Enzyme, hormone, antibody, structural, receptor **any one of these**	1	There are five different possibilities to learn.

Question			Expected response	Mark	Top Tips
4.	(a)		To ensure that both solutions were at the required temperature **OR** To ensure that the reaction took place at the correct temperature.	1	Good experimental procedure and one to remember.
	(b)		Same volume of lipid/lipase Same concentration of lipid/ lipase Same pH **Any 1**	1	Notice solutions in experiments and for variables think of **V**olumes and **C**oncentrations- **VC**
	(c)		As the temperature increases the activity of the lipase increases up to 40 °C and then decreases.	1	Be careful to refer to 'the activity of the lipase' and make sure this term is used. Also make sure that you have described the full range of temperatures.
	(d)		Repeat the experiment at each different temperature	1	Make sure you mention the need to repeat at **each** temperature.
5.	(a)		Testes	1	You need to know the name of the organs that produce the gametes in **both** plants and animals. Only animals needed here, though.
	(b)		Provide energy/ATP required for sperm to swim towards egg	1	Your answer must include reference to the 'large' number of mitochondria and requires you to explain that sperm are very active and require lots of ATP/energy to swim towards the egg.
	(c)		Contains one set of chromosomes	1	Good definitions and terminology will help you nail these questions. Add them to your 'Flash card' pack of key words and phrases.
	(d)		Sperm nucleus and egg nucleus fuse or join together/sperm and egg nuclei fuse/gamete nuclei fuse or join together	1	Notice that the answer requires you to specify that the **nucleus** of the sperm and egg fuse together. It was not enough to say that 'the sperm and egg fuse together'.

Question			Expected response	Mark	Top Tips
6.	(a)		Scales and labels = **1** Plots and connection = **1**	2	Include zeros and highest values on even scales. Include units with labels. Plot with a sharp pencil. Connect plots with straight lines.
	(b)		5:4	1	Make sure you have whole numbers that do not have a common factor.
	(c)		Alcohol is toxic to the yeast cells **OR** glucose all used up	1	Standard answer to this question.
	(d)		Use a finer scale/narrower container/scale in mm^3 to measure CO_2 volumes **OR** use a CO_2 probe and data-logger	1	Again, accuracy is related to measurements.
	(e)		Repeat but vary the chosen factor and keep all other factors the same	1	Three steps as usual. Repeat, original factor held constant, chosen factor varying.
7.	(a)		Electrical impulses	1	Nervous messages are electrical impulses.
	(b)		Release of chemical into gap/ synapse		The electrical impulse can be carried through the gap by chemicals.
	(c)		Synapse	1	The gaps prevent continuous transmission of nerve impulse – you just have to learn the name.
	(d)		Automatic/rapid = **1** Protects body from harm = **1**	2	The rate of response prevents a harmful stimulus from causing damage.
8.	(a)		1.4 cm	1	Subtract the shortest from the longest.
	(b)		There is a range of lengths involved	1	Remember that continuous variation shows a range of values that merge with each other – the values are not clear-cut.
	(c)		Correct matching example, e.g. human tongue-rolling; rose-petal colour	1	It's useful to have a few examples ready – human ones are often easiest to remember.

Question			Expected response	Mark	Top Tips
9.	(a)	(i)	P on left upper vessel on diagram	1	You must learn the pattern – the left atrium position is the reference clue.
	(a)	(ii)	Downward arrow	1	Remember the left side is on the right of the diagram! Think about your reflection in a mirror.
	(b)		Valves	1	Any type of valve permits one-way flow only.
	(c)		Have a biconcave shape which gives a large surface area = **1** Have no nucleus which allows more space for haemoglobin = **1** Haemoglobin transports oxygen (as oxyhaemoglobin) = **1** **[Any 2]**	2	A large surface area is useful in any absorbing surface and the presence of haemoglobin allows the transport of oxygen.
10.	(a)		7 hours until 18 hours	1	Draw a line across the graph from 100 cm^3 per hour per plant then read down from the intersects to the times on the scale.
	(b)		1: temperature = **1** 2: light intensity = **1**	2	Looking for environmental factors that vary through a day – these are really the only options. Wind could be involved but it is not a predictable feature of a day.
	(c)		Stoma/stomata	1	Remember the pore is the stoma that is formed between the guard cells.
	(d)		Carries water to leaves for photosynthesis **OR** cools plant **OR** provides support for cells	1	The standard benefits of transpiration!

Question			Expected response	Mark	Top Tips
11.	(a)		Quadrats = **1** Drop randomly and count/record numbers of plants inside = **1**	2	Quadrats are the only sampling method for plants mentioned in National 5 assessment. Very simple to use.
	(b)		205	1	Go to the path chart and species 1 – every small box is five plants so 205!
	(c)		Species O – decreases in number Species P – increases in number	1 1	Bits of different highlighter colour on the species columns of each graph might help.
	(d)		Species Q	1	Species Q stands out as unchanged – in some questions you might have to read the bars carefully.
12	(a)		Woodlice-10 segments Snails – 1 segment Centipedes – 3 segments Beetles – 7 segments **Segments correct = 1** **Labelled/keyed correctly = 1**	2	Practise constructing similar pie charts and remember to label them or use a key.
	(b)		Precaution-Level/flush with soil surface; no gaps; cover or lid; others = **1** Improvement – Repeat/use many traps = **1**	2	Make sure that you have learned some of the sources of error for both sampling techniques and measurement of abiotic factors.
13.	(a)		8000	1	Like the limiting factors in photosynthesis, the factor on the X-axis is only limiting when the graph is still rising. Once the graph levels off, it is no longer a limiting factor.
	(b)		300	1	Percentage increase = Increase/Starting value × 100 So, Increase = 8000 – 2000 = 6000 6000/2000 × 100 = 300
	(c)		Protein, polypeptide, amino acid, nucleic acid	1	Plants take up nitrate, convert it to amino acids, which they use in protein synthesis.

Question			Expected response	Mark	Top Tips
	(d)		Fertilisers	1	Fertilisers usually provide NPK – nitrogen, phosphorus and potassium.
14.	(a)		False True True **All 3 for 2 marks, 2 or 1 for 1 mark**	2	Remember **ROALF** – **R**andom **O**ccurrence **A**nd **L**ow **F**requency.
	(b)		Organisms vary Best adapted varieties have a selective advantage These varieties survive better They pass on their genes to their offspring **All 4 for 3 marks, 3 for 2 marks, 2 for 1 mark**	3	Natural selection acts on variation – this is the basis of evolution.
15.	(a)		The further away from the source of pollution, the more lichen cover	1	What do the data show? Is there a trend? Does the trend relate to the statement?
	(b)		8 km	1	Find the high point of the data and read down to the scale.
	(c)		Reduce photosynthesis = **1** Blockage of light **OR** clogged stomata = **1**	2	Think about sooty dust and leaves – How could dust affect photosynthesis?
16.	This type of question is designed to test both skills and knowledge. Read these short passage questions carefully. It is sometimes quite useful to have a quick skim through the questions being asked before reading the passage. For most answers you can give the same phrases used in the actual passage, unless asked for a particular meaning or explanation.				
	(a)		Twice	1	Difference for lying down = 16; difference for running = 32 so **twice** 16 = 32
	(b)		More oxygen is required for more respiration to produce more ATP/energy	1	Bigger decrease in oxygen saturation for increased activity which shows more oxygen required.

Question			Expected response	Mark	Top Tips
	(c)		Thick wall in arteries, thinner wall in veins **OR** wider diameter in arteries, narrower diameter in veins **OR** veins contain valves, arteries do not	1	Answer from your own knowledge here.
	(d)		Capillaries	1	

Practice Paper C

Section 1

Question	Response	Mark	Top Tips
1.	D	1	More concentrated in solutes means lower concentration of water so water moves out – that causes the cells to shrink.
2.	A	1	The **p**atchy molecules in the membrane are the **p**roteins.
3.	A	1	Look for the concentration gradient which matches the answers.
4.	C	1	Bases are complementary: A always pairs with T, and G always pairs with C.
5.	B	1	Remember, the pH scale measures acidity – the lower the pH number, the more acidic. Look for the peaks at the low pH numbers.
6.	C	1	Remember all respiration starts in the cytoplasm but only aerobic requires further breakdown in the mitochondria.
7.	A	1	Pyruvate is always the start point and in animals the produce is lactate.
8.	B	1	**All** stem cells self re-new and can become specialised.
9.	D	1	Straight from the Course Specification.
10.	C	1	Remember gluca**gon** is needed when glucose is **gon**e.
11.	C	1	Find the increase, divide it by the start point then multiply by 100

Question	Response	Mark	Top Tips
12.	D	1	Flowers are different shapes but the internal parts are always laid out in a similar way.
13.	D	1	If a parental characteristic does not show in any of its offspring, then its allele must be recessive.
14.	A	1	Very typical drawings.
15.	B	1	Lignin occurs in rings and spirals and is very tough material.
16.	D	1	Link villi with the small intestine then pick out the small intestine in the diagram.
17.	B	1	Why not make some flash cards of the terms in this question? The terms come up a lot and can be confusing.
18.	C	1	Look at the clues in the drawing.
19.	A	1	This is an unusual pyramid and obviously starts with a big plant. The large top often indicates parasites.
20.	A	1	The answer is based on knowing that intraspecific means within the same species. The herring are different stages of the same species.
21.	C	1	The energy in ATP is chemical but it is the energy in light that is required to make it. Water (H_2O) is split into hydrogen and oxygen.
22.	A	1	When the graph slopes, the limiting factor is on the x axis. If the graph line is flat, another factor is limiting – notice the effect of increased temperature at R.
23.	D	1	This can be done by eye but it's better to work out the ratio for each year – this won't take long because the light form is the same each year.
24.	B	1	Note that nitrates are used to produce amino acids but that the amino acids might then be used later to produce protein.
25.	B	1	Quite easy if you are careful and read the information given.

Practice Paper C

Section 2

Question			Expected response	Mark	Top Tips
1.	(a)		Chloroplasts are present	1	Chloroplasts are the sign of the ability to photosynthesise.
	(b)		Aerobic respiration **OR** production of ATP	1	You have to recognise the mitochondria and know their function.
	(c)		Ribosomes	1	Remember that ribosomes can be attached to a membrane or be free in the cytoplasm.
	(d)		Walls composed of different substances **OR** fungal cells don't have chloroplasts	1	You need to know about differences in cell walls and that fungi don't have chloroplasts.
2.	(a)		Active transport = **1** Low to high concentration **OR** transported by membrane protein = **1**	2	The concentration gradient shown by the molecules in the diagram is the clue!
	(b)	(i)	Scales and labels = **1** Plots and joining = **1**	2	As usual, remember the basic points.
	(b)	(ii)	Osmosis	1	When cells take up water they gain mass.
	(b)	(iii)	1.8 g	1	Get the % increase from the table (90%) then apply it to 2.0 = 1.8 g.
3.	(a)		DNA	1	Genes are made up of DNA (apart from some virus genes, which are RNA).
	(b)		Plasmid	1	Plasmids can be removed, altered, then put back into bacterial species.
	(c)		To extract required gene from chromosome/to insert required gene into plasmid	1	Since enzymes can degrade and synthesise, you are looking for parts of the process where these processes might be used.
	(d)		Insulin **OR** growth hormone	1	There are many others but it's probably better to learn these ones.

Question			Expected response	Mark	Top Tips
4.	(a)		8%	1	A two-part calculation: first, take the actual decrease and divide it by the starting length; second, multiply the answer by 100. (4 /50) x 100 = 8%
	(b)		Energy in glucose has not been released = **1** ATP is a source of instant energy = **1**	2	The energy in glucose must be released by respiration – this can only happen in live tissue!
	(c)		As a control to show the effects of glucose and ATP	1	The control shows that the factor that was causing the contraction in muscle B was the ATP not the water, which would have been present in the ATP solution.
	(d)		A group of similar cells carrying out the same function	1	Remember the sequence: Organelle – cell – tissue – organ – system – organism.
5.	(a)		DNA	1	Chromosomes are made of DNA. Sections of DNA are called genes.
	(b)		Spindle fibres pull chromatids apart = **1** Chromatids/chromosomes move to the ends/poles of the cell = **1**	2	Writing each step in the sequence onto cards then practicing putting them into order is a good activity here.
	(c)		Growth **OR** repair	1	These are the uses mentioned in the Course specification.
6.	(a)		A: sensory neuron B: inter neuron C: motor neuron **All 3 for 2 marks, 2 or 1 for 1 mark**	2	The names of the neurons are clues to their functions.
	(b)		Heat = **1** (Muscular) withdrawal movement = **1**	2	Reflexes are protective so the stimulus is potentially damaging.
	(c)		Rapid response provides protection	1	Response has to be rapid because the stimulus will be starting damage immediately. Try **RAP** = **R**apid, **A**utomatic, **P**rotective.

Question			Expected response	Mark	Top Tips
7.	(a)		Temperature Light intensity Watering **2 marks for any 2 of these,** **1 mark for only one**	2	These are the very basic variables for plant growth.
	(b)		43–63 cm **OR** 20 cm	1	The range is the smallest in the sample through to the largest in the sample and can also be expressed as the difference.
	(c)		Gap between tall range and dwarf range caused by genetic differences = **1** Range within a variety = **1**	2	In discrete variation there are clear-cut differences – here the cut comes between the tallest dwarf plant and the smallest tall plant!
8.	(a)	(i)	dd	1	Female parent has no dimples which is the recessive characteristic and can only be dd.
	(a)	(ii)	Dd	1	Son R has dimples so must have at least one of these dominant alleles D. However he has to inherit a recessive d from his mother.
	(b)	(i)	50%	1	Like Son R, Daughter Q has a Dd genotype. Her male partner has no dimples and must be dd. Use this to draw a Punnett square.
	(b)	(ii)	Fertilisation/fusion of gametes is a random/chance process	1	Learn this phrase to explain why actual ratios can differ from predicted ratios.

Question			Expected response	Mark	Top Tips
9.	(a)	(i)	X: vena cava = **1** Y: coronary artery/vein/vessel = **1**	2	The clue for X is the direction of blood flow. For Y it is the fact that the vessel is attached to the outer surface of the heart.
	(a)	(ii)	True False capillaries False veins **All 3 for 2 marks, 2 or 1 for 1 mark**	2	**A**rteries carry blood **A**way. **V**eins have **V**alves.
	(b)	(i)	0.4 litres	1	The 2% has to be applied to the 20 litres inhaled to get the carbon dioxide.
	(b)	(ii)	38 litres	1	For this question, drawing lines on the graph using a ruler will help avoid misreads.
10.	(a)		1: mass of water lost as shown on balance 2: area of leaf to be measured 3: time to be measured **1 each**	3	You need to look at the units of transpiration on the table and think how each part would be measured.
	(b)		Repeat the investigation but… alter chosen factor = **1** keep all other factors same = **1**	2	Standard three-step approach needed again – repeat the experiment, keep the original variable constant and vary your chosen factor.

Question			Expected response	Mark	Top Tips
11.	(a)	(i)	The habitat and the non-living components/abiotic factors	1	Learn the definition of an ecosystem. Flash cards and matching exercises are really useful to help you remember the large number of definitions you need to know.
	(a)	(ii)	An inherited characteristic that makes an organism well suited to survival in its environment/niche	1	This is the definition straight from the course specification-best use it.
	(b)		38%	1	Just add up the percentages of the different forest-based ecosystems.
	(c)		Role played by an organism	1	A difficult idea so remember the word *role* as an alternative.
	(d)	1	Sunlight/light energy is trapped by chlorophyll/chloroplasts **and** converted to chemical energy/ATP	1	Photosynthesis is energy conversion from light to chemical
		2	Lost as heat/by movement/in undigested material	1	These are the three fates listed in the course specification
12.	(a)		Number of traps used Method of setting Same time of day Time left **1 each for any 2**	2	It's really common sense to keep the sampling method the same so that the method does not influence the sample differently for different areas of grassland.
	(b)		Not enough traps Not randomly set Badly set Left too long before checking **any of these = 1**	1	The sources of error are not so much with the traps as with the methods of using them! For example, if the trap lip is above the soil surface or if predators are able to consume the sample!
	(c)		Moisture, temperature, pH = **1** Moisture meter, thermometer, pH meter/paper = **1**	2	Remember that the instruments are all **meters**.

Question			Expected response	Mark	Top Tips
13.	(a)	(i)	Oxygen	1	Remember oxygen is a product of photosynthesis.
	(a)	(ii)	Change the light intensity with a dimmer/by moving the lamp nearer or further from the plant = **1** Count bubbles/measure gas production = **1** Over a set period of time = **1**	3	This method of measuring the rate of photosynthesis is one of the seven experimental techniques you must know about in detail.
	(b)	(i)	0.9–0.95%	1	Find the maximum rate at 15°C then use a ruler to draw a line down to the X axis and read off the value - note that **2** small divisions = 0.1%.
	(b)	(ii)	0.05 units	1	Draw a line up the grid from 2% until it cuts both graphs. Calculate the difference in rates between the graph lines (0.2 – 0.15 = 0.05).
	(c)		Storage	1	Remember the **st**arch is for **st**orage!
14.	This type of question is designed to test both skills and knowledge. Read these short passage questions carefully. It is sometimes quite useful to have a quick skim through the questions being asked before reading the passage. For most answers you can give the same phrases used in the actual passage, unless asked for a particular meaning or explanation.				
	(a)		Plants are inefficient at taking it up	1	Straight from information in the passage.
	(b)		They become food for oxygen consuming bacteria	1	Straight from information in the passage.
	(c)		245.5 metric tonnes	1	Just subtract the appropriate figures from the passage.
	(d)		GM rice is more efficient at taking up nitrate from soils	1	Straight from information in the passage.
	(e)		Higher yields **OR** lower costs **OR** better food security	1	Straight from information in the passage.

Question			Expected response	Mark	Top Tips
15.	(a)		Isolated populations of finches = **1** different mutations occur in different populations = **1** natural selection ensures only best suited organisms survive to breed = **1**	3	What could be stopping interbreeding occurring? It's important to say that mutations are **different.** It's survival to **breed** which is crucial to evolution.
	(b)		Neutral	1	Some mutations have little apparent effect on the organism – they are neutral – neither good nor bad.